Nature and National Identity
after Communism

Pitt Series in Russian and East European Studies

JONATIIAN IIARRIS, EDITOR

Nature and National Identity after Communism

Globalizing the Ethnoscape

KATRINA Z. S. SCHWARTZ

UNIVERSITY OF PITTSBURGH PRESS

Published by the University of Pittsburgh Press,
Pittsburgh, Pa., 15260

Copyright © 2006, University of Pittsburgh Press
All rights reserved
Manufactured in the United States of America
Printed on acid-free paper
10 9 8 7 6 5 4 3 2 1

Library of Congress Cataloging-in-Publication Data
Schwartz, Katrina Z. S.
    Nature and national identity after communism : globalizing the ethnoscape /
Katrina Z. S. Schwartz.
        p. cm. — (Pitt series in Russian and East European studies)
    Includes bibliographical references and index.
    ISBN 0-8229-4296-8 (cloth : alk. paper) — ISBN 0-8229-5942-9 (pbk. : alk. paper)
        1. Human ecology—Political aspects—Latvia. 2. Political ecology—Latvia.
3. Landscape ecology—Latvia. 4. Environmental policy—Latvia. 5. Environmental
protection—Latvia. 6. Conservation of natural resources—Latvia. 7. Rural development—Latvia. 8. Latvia—Environmental conditions. 9. Latvia—Politics and government. I. Title. II. Series.
    GF602.4S34 2006
    333.72094796—dc22   2006015726

Versions of chapters 2, 5, 6, and 7 have been published as: "'The Occupation of
Beauty': Imagining Nature and Nation in Latvia," *East European Politics and Societies*
(forthcoming); "'Masters in Our Native Place': The Politics of National Parks on the
Road from Communism to 'Europe,'" *Political Geography* 25, no. 1 (January 2006):
42–71; "Wild Horses in a 'European Wilderness': Imagining Sustainable Development in the Post-Soviet Countryside," *Cultural Geographies* 12, no. 3 (2005): 292–320;
and "Nature, Development, and National Identity: The Battle over Sustainable
Forestry in Latvia," *Environmental Politics* 8, no. 3 (Autumn 1999): 99–118.

In loving memory of
Laura Zinaīda Strautzels (1905–2003),
Strautzeļu saimniece

CONTENTS

ILLUSTRATIONS

ACKNOWLEDGMENTS

I would like to thank, first of all, Kathryn Hendley, Mark Beissinger, and Robert Kaiser at the University of Wisconsin-Madison, whose expertise, disciplinary tolerance, and unflagging support enabled me to weave together the hybrid project from which this book emerged. A number of people read and commented upon some or all of this manuscript. Thanks are due to Jane Dawson; Aldis Purs; Catherine Wanner; and anonymous reviewers for *Cultural Geographies*, *East European Politics & Societies*, *Environmental Politics*, *Political Geography*, and the University of Pittsburgh Press; as well as my editor, Peter Kracht. This book would have been infinitely less polished and rigorous without the extraordinary guidance of my delightful postdoctoral group at the Harriman Institute: Hilary Appel, Victoria Frede, Rebecca Neary, Ethan Pollock (surely the best style editor I have ever had the good fortune to learn from), and Olga Shevchenko.

The research for this book was conducted primarily during a year of fieldwork in Latvia in 1998–1999, which was supported by a Fulbright-Hays Fellowship for Doctoral Dissertation Research Abroad, an IREX Pre-Doctoral Fellowship and a University of Wisconsin Vilas Travel Fellowship. Write-up support was provided by a MacArthur Global Studies Fellowship, an ACLS East European Studies Dissertation Fellowship, and a postdoctoral fellowship at Columbia University's Harriman Institute.

I am deeply grateful to the many individuals in Latvia who shared their time, thoughts, and passions with me; the list is too long to enumerate here. I thank Katrīna and Jānis Pērkons of Rucava for their outstanding hospitality, exquisite anecdotes, and boundless warmth and good will; the same goes for Zaiga and the other "Kates" of Vērgale. Māra Zirnīte and her associates at the Latvian Academy of Sciences Institute of Literature, Folklore, and Art earned my eternal gratitude for their faultless transcribing, as well as for some fascinating conversations high up in the Stalin tower. I am grateful to Aija Zviedre for providing easy access to the collection of Latvian forestry periodicals at the Latvian Forest Employees' Association. An enormous thank-you goes to Guntis Eniņš for sharing with me his vast personal archive of Soviet-era photographs and news clippings concerning the Great Tree Liberation Movement and for consenting to have some of those photographs reproduced here. I thank the entire staff of WWF-Latvia for their technical support and for suffering cheerfully their fate as my guinea pigs, but mostly for the friendship and the many, many laughs I enjoyed at my home away from home on Krišjānis Barons Street and later on Elizabetes. The happy hours, Internet pranks, and *Fawlty Towers* screenings made the time pass so much more easily. Most of all, thanks to Uģis Rotbergs for sharing unstintingly his encyclopedic knowledge and challenging analyses, for his humor, and for the example of his unflagging dedication to nature and nation.

My fieldwork year was made much more enjoyable by the good times shared with several dear friends, especially Inese Lapa, Mārtiņš Hildebrants, Astrīda Lēvenšteina, Mārtiņš Parņickis, Ilva Cimdiņa, and my sister Māra Schwartz (now Kore). Māra's role as my "spy" at Lake Pape was a great boon in terms of both research and recreation, and her presence took the edge off of the loneliness of fieldwork. Long live the condiment revolution! A huge thanks to my mother, Dr. Ilze Knezinska Schwartz, for providing the deluxe living quarters in downtown Riga that, by some accounts, disqualified that year as a genuine fieldwork experience.

A number of people have provided inspiration and opened up new worlds to me. I would like to thank Margaret Corey, wherever she is, for an unforgettable year with Herman Melville; Liene Dindone and Lalita Muižniece for easing my entry into the sometimes unwelcoming world of Latvian-American society; and William A. Zeps for teaching me so much about love of landscape. My years in Madison were blessed by the presence of my second cousin, Agate Nesaule, whose beautiful and moving memoir *A Woman in Amber* is an inspiration, and whose warmth and love (along with the all too infrequent visits from her sister Beate Kukainis) buoyed my spirits tremendously during the lonely travails of graduate school.

Last but not least, I am deeply grateful to my family: my sister Anna for fearlessly (and always lovingly) blazing the trail for her more meandering little sister; my wonderful grandfathers, the late Earl Schwartz and Jēkabs Knezinskis; and especially my late grandmother, Zenta Knezinskis, and my parents, Ilze and Richard Schwartz, for their unbridled enthusiasm for my work and their generous material and moral support. Most of all, I owe everything to the boundless love and wisdom of my partner, Sue O'Brien, who has made it all worthwhile—and who always makes me laugh along the way. Without her, I would never have enjoyed the exuberant love and companionship of Nellie, Jessie, Roxanne, and the late great Toby and Talley.

This book may well have never come to be without the childhood years spent at the kitchen table with my great-aunt Laura Strautzels (1905–2003), a remarkable storyteller with a joyous love of words in four languages (Latvian, German, Russian, and English, plus the occasional Ovid fragment recalled from high school). Aunt Laura's photographic memory, wry humor, keen intelligence, and elegant narrative style brought to life for me the vanished landscapes of pre-Soviet Europe, through her accounts of life on our family's prosperous farmstead, as a university student in Riga and a provincial schoolteacher during the interwar First Republic, and during two world wars and two exiles (the second to America; the first to Tver in Russia, where she lost a beloved little sister, made friends with Russian classmates, and rose to first in her class at the Rimskaya-Korsakova School for Girls). Because of Aunt Laura, the past has always seemed to me like a foreign country still open to visitors.

When I began working on this project, I talked about it with Aunt Laura during my visits to the family home in Cleveland. As always, she listened with acute interest. At some point she noted that my research themes made her think about the family farmstead, Strautzeļas, where she grew up and to which she returned often in adulthood as the de facto head of household, taking over for her less practically inclined mother. I asked her to tell me more, but she said she needed to think about it. The following morning she announced that she was ready to talk, and I sat down with my tape recorder. I ask the reader's indulgence in reproducing here the narrative she delivered to me on that September morning in 1997, not only because it will serve as the only published testimony to her astonishing memory and verbal gifts, but also because it encapsulates so many of the themes of this book. (As always, alas, much of the beauty is lost in translation.) Like the agrarian memoirs discussed in chapter 1, her story begins with idyllic childhood memories of the Strauzeļas farmstead, but it concludes with the despoliation of her own personal ethnoscape by the forces of wetland drainage (*meliorācija*) during the late 1930s, at the zenith of agrarian nationalist productivism and on the eve of World

War II and the Soviet occupation of Latvia. "Do you know what title I want to give it?" Aunt Laura asked me before she began narrating. "The ditch."

★ ★ ★

Strautzeļas was a farmstead. The fields were primarily for grain. All the fields were divided into geometric shapes, squares or triangles or some other shape, with the help of ditches. And these ditches were the only drainage. This drainage was very primitive. The fields were a little bit sloped, and there was a sort of low-lying area. That was left as an uncultivated meadow, and it stretched the entire length of the property along that edge—about two kilometers. No one fertilized it or mowed it, there was no hay—they just kept it. And a little stream flowed through the meadow, a little meandering stream. And on the banks of the stream were tree stands in two places. We had the Big Grove and the Little Grove. And funnily enough these groves formed themselves—very little work had to be done in them. We all very much loved the near one—not just the children, but the entire household loved it, and if people came from Riga, or sometimes guests from Jelgava, then we would eat breakfast and—"Let's go to the grove."

The biggest event for the Strautzeļas household and for this stream was springtime. My, how we awaited the spring! It occurs to me to compare it to provincial towns in America awaiting the arrival of a touring performance by the Metropolitan Opera. The performances run for maybe two weeks. And everyone waits nervously before it, and enthuses after they've left, and then there is something to talk about and remember. That's how at Strautzeļas every year you awaited the same old springtime. Every year. Everyone—not just the children. The children most of all, but also for example the farmhands and herders—everyone went to check the pond. (There was also a pond in the meadow, and a well and a sauna, and also one farmhands' house, where the laborers could live.) And you went to see whether spring had arrived, so as not to miss the very beginning. First the pond thawed, full of frogs croaking. Then you looked—yes, the eggs have been laid. The eggs weren't one by one, but in clumps that were called spawn. And the children went to watch how the spawn grew. But there would still be snow here and there. Suddenly the stream swelled—it rose up and was no longer calmly blanketed with its layer of winter snow, but began to split, and water began to appear here and there, and then it jostled, and the snow also started to push around. Well, and that was the culmination.

Then the "visitors" began to arrive—the songbirds. That was called the nesting time, and they all had it at around the same time. The starlings had already arrived before them. They lived in little man-made wooden houses, attached to apple trees in the orchard, which were inspected before the spring arrival to see if they were still in good shape, or whether a new little board needed to be attached for the bird to sit on when he sings in front of his house. And the starlings began to whistle.

It sounds like bells in the sky—you look: blue skies, you don't see anything, but everything is ringing, ringing, it sounds like either a stringed instrument, or silver bells. What is it? The skylarks. They hover as high as they can in the sky. But why do they sing, and why do they hover right there, above the gray field? There is nothing on the field, or maybe there is a thin green layer of winter grain (rye and wheat were planted in the fall). They hover up there and sing because down in that bare field is the mother on her eggs. My, how beautiful those skylarks are!

The biggest tree in the grove, which stood apart from the others, had its top cut off and a wheel had been attached horizontally, and that's where the storks were awaited. They came in pairs, sometimes two pairs, sometimes only one. When the storks arrived, they tested the wheel themselves, jumping on the rim to see if it held.

And then also all sorts of little birds. Once they arrived, then quite soon the whole land, the whole grove, and the whole meadow suddenly was one great carpet of flowers—exactly like a carpet. In the low places, right by the banks of the stream, there were yellow marsh marigolds. Everything was carpeted with marsh marigolds. In the shadier places were anemone flowers, also yellow. Then higher up were yellow primroses and purple swallow-wort. And it all looked like a carpet, throughout the entire meadow.

There were some old trees, and they had lots of hollow places. Birds nested in all the hollow trees. And we were taught that you could look, but you couldn't breathe on them, and you couldn't show your teeth or talk: then the bird would abandon the nest. And we three sisters would go from tree to tree—we knew where the old trees were. What were these old trees? They were rowan trees and crab apple trees. And the branches were knotted, and often water would gather in the knots, and there the birds had made their nests. We lifted each other up as best as we could in order to see, because the nest was a little too high for us.

The little grove was especially beautiful on this side, where we brought guests. It was mostly birches. And under the birches almost nothing grew on its own, except for the junipers. The birches—well, it was a park, just like a park. The birches would already have leaves. A little farther along the stream bank, where the birches ended, were hazelnut trees, which let out a spray of golden pollen when they blossomed. Then with the birches there were also rowans, and another wonderful tree—the bird-cherry, and that was blossoming too. The nut trees first.

And even before the nut trees, the pussy willow catkins—already in late February they were bright silver, and the branch was a sort of bright brown color.

So the storks were making their nest, and the meadow was full of little baby frogs, and they all had food, and all the birds had food. And there were all sorts of insects, sort of a natural jungle, and they had good pastures there. When the bird-cherries had started to bud out, one lovely evening the nightingales began to sing. Wonderful! They would sing all night. They say about the nightingale that when she's singing she suddenly loses her voice, and you think she's gotten tired of it. No! She gets so carried away that sometimes she has a heart attack—a heart attack from ecstasy. The mother hatches the eggs, the father sings. All night until the light of dawn, and the bird-cherries are getting ready to bud out, to blossom.

And that was the "touring performance," and it lasted around a month. The storks had arrived from Africa, where they had been waiting to return to their homeland. The majority of these birds came back to their place of birth. Sometimes the very same storks come and come—suddenly one comes back alone. His partner had probably died on the journey. Sometimes he lives alone, he can't get a new partner. He lives out the summer alone in the nest, without children. . . .

But, well, life progresses, doesn't it? [As an adult] I start to get called in to the township council. "We have a project and you must participate in the project. We want to drain the fields, and you must give your one-and-a-half-kilometer stream in your meadow, so that a drainage ditch can be dug there. And you must join the drainage society." I refuse. They ask me why, and I say, "That is the most beautiful place," and I tell them about it. I say—when visitors come from Riga and other cities, then we go to the grove and everyone likes it there very much, and it's beautiful there in late summer too. "That's unacceptable," they say, and, "Nowadays everyone is draining, and everyone needs that drainage ditch, and you and your farmstead are backward, and the way that grove looks—I'm ashamed to drive by it!" I don't listen, and for a good long while I don't listen.

Well, finally I have to listen after all. I agreed to join the drainage society. That's how it is—progress, everyone wants artificial fertilizers, everyone needs the drainage ditch, and they all run their ditches into the big ditch. And I saw how the ditch was dug; it was around two meters wide, fairly deep. And so it happened.

And it's funny—from that very summer we notice that we no longer have flowers, that nothing grows there anymore. Everything starts to change. And here and there a tree dies. And we are surprised to see owls showing up in the apple orchard. One time, for example, we were sleeping in the attic room, the windows were opened outward, and I see on the wall something like the shadow of a bottle. I turn my head to see what it could be—an owl is looking right at me! An owl coming into the bedroom! Another time I go to the window downstairs to see what's shining there. Two owls in the cherry orchard. It was strange—owls started coming to the house. They lived in the woods, in the groves. We noticed them for several summers. And then the nightingales started singing right in the yard. And they sang all night. We kept the windows open at night, and one night my mother awakens me and says, "Can you sleep? I can't sleep when the nightingale comes so close to sing." It had come right up to her window.

So we started wondering, what is this? At first we didn't really notice the changes—well, only that the ditch was ugly. Only that much. And then everyone began to talk about how the grove dwellers were coming to our yard at night, as if to complain. And we talked about it and decided—something is going to happen, something terrible. In the old days, when forest birds and forest animals come to people's houses, that means war is coming.

Well, and that's how it ended. And all this time I still thought about the ditch and the changes in that grove, and the birds' strange behavior, when they began staying in the yard right by the windows—I identified it with a prophecy of disaster from God. I was convinced that the old folks were right! That if a forest owl comes to the house, it means war. Until here [in America] I started hearing about ecology, about the Amazon River, and how they cleared the forests improperly there, and so on. It occurred to me that the change in that grove, and everything that happened, was ecology. At that time I didn't know about this ecology. And so our "wetland" was transformed into nothing. Into nothing. It had to be drained and planted with wheat and rye. And all around there wasn't a single grove—ours had been the only one. Everyone was draining, and it kept getting worse and worse in that grove, and one after another the trees died. And then in the Communist times they rode over it with bulldozers and flattened the grove.

Nature and National Identity
after Communism

Introduction

On May 1, 2004, eight formerly Communist countries in Central and Eastern Europe, including the three former Soviet republics of Latvia, Lithuania, and Estonia, consummated their long-desired "return to Europe" by joining the European Union (EU).[1] The new members brought with them well-known problems of poverty, inequality, corruption, and popular disillusionment with democracy and the market. Many of these problems are particularly severe in rural areas, where nearly 4 million farmers will now compete in the European common market with the 6.5 million agriculturalists of existing member states. The region brought with it also the scars of environmental damage wrought by decades of state-socialist industrialization: the lingering effects of the Chernobyl disaster, forests laid waste by acid rain, the old strip mines and coal-fired power plants in the so-called black triangle on the East German–Czech–Polish border.

While popular perceptions of the East European environment center on nightmarish images of pollution hotspots, there is far less recognition of the fact that Communism yielded profound, if often unintended, environmental benefits as well: namely, protection for a wealth of relatively untouched nature that supports far more biological diversity than can be found in Western Europe.[2] This wealth testifies less to the superiority of the Communist states' toothless environmental protection laws than to the relative ineffectiveness of state-socialist economies in achieving the goal that Marxist-Leninist ideology shared with its capitalist nemesis: transforming nature for productive purposes. Part of the explanation, nevertheless, lies in the impressive system of protected nature areas established throughout the region during the Communist period and perhaps also in the presence of a strong, historically rooted popular environmental ethos.[3] But whether by accident or design, in May 2004 not only were 74 million human survivors of Communism welcomed into the European fold: so too were thousands of brown bear, wolves, and lynx roaming the Carpathian Mountains and the world's densest population of black storks nesting secretively in Latvia's "northern rain forests." The coal mines of Bohemia have been "returned to Europe," but so have Poland's "primeval" Bialowieza Forest, the flood plains of the Danube, and the high dunes of Lithuania's Couronian Spit.

An abundance of nature is one face of an unexpected dual legacy bequeathed to the post-Communist countryside when the Soviet Union and its satellite regimes collapsed. The other is a relatively oversized and inefficient agrarian sector, which throughout the region employs a far greater share of the population far less productively than in advanced capitalist industrial societies. The elimination of Communist-era subsidies and trading blocs has ravaged agricultural livelihoods, and future prospects for these farmers are dubious at best; Europe is already burdened with agricultural overcapacity and is struggling desperately to reform and reduce its own much-criticized system of agricultural supports.[4] But while the withering away of the agricultural sphere is likely to continue and indeed accelerate in these countries, demand for the East's natural wealth is likely to grow in the crowded cities and manicured suburbs of postindustrial Western Europe.

As one analyst of the geography of post-Communist Europe puts it: "Directly or indirectly, in greater or lesser measure, the lives of 400 million people (or about 8% of world population) are being reshaped; the functions, values and perception of space across the 17% of the earth's landmass that they occupy are being reorganized, reassessed and redefined."[5] The Central and East European countries confront a developmental challenge that will have profound consequences not only for the livelihoods of millions of people but also for natural diversity on the European continent. What is to be done with large and underemployed rural populations and with growing areas of marginal farmland? Can nature itself be used as a resource to promote rural development? And if so, with what consequences, both for nature and for people? Responses to this challenge will evolve during the coming years, and they will be informed by many factors: by the give-and-take of national politics; the specific conditions of the environmental, agricultural, and rural development assistance dispensed by EU and international donors; the vagaries of global market demand; and the strictures of market liberalization. They will also be informed by the ideas and passions of post-Communist rural dwellers themselves: by how they understand the meaning of rural landscapes, the value of agrarian labor, and the proper relationship between human beings and nature.

This book hopes to shed light on this last dimension by exploring local responses to Western aid initiatives in nature management and rural development in one post-Communist country, Latvia, during the 1990s (that is, before its accession to the EU). Debates about the future of the Latvian countryside have been deeply influenced by competing understandings of nature, and these, in turn, are intimately intertwined with discourses of nation. To understand how Latvians have negotiated and will continue to negotiate their "return to Europe" and, more broadly, their reinsertion into the cultural and economic space of the West, we must understand how the uses and meanings of nature are bound up with national identity: how generations of Latvians have defined Latvianness in terms of a particular relationship to nature and have constructed nature as a particular embodiment of Latvianness. The nexus of nature and nation is, of course, by no means unique to Latvia, and I present the Latvian case as one post-Communist variation, with some distinctive elements, on a universal theme. Like others elsewhere in the region and in other places and times, many Latvians believe that what is at stake in the incipient Europeanization of the countryside is the fate not only of storks and wolves and wetlands, but also, as one rural dweller told me, of "Latvians who know they are Latvians."

Constructing Nature and Nation

"Nature," as William Cronon puts it, "is not nearly as natural as it seems. Instead, it is a profoundly human construction." Social constructions of nature reveal much about "not just the natural world but the human cultures that lend meaning and moral imperatives to that world."[6] Among other things, they reveal much about conceptions of national identity, for the cultural entrepreneurs of nations great and small are forever seeing the soul of the nation reflected in its native landscapes. Indeed, the perceived symbiosis between a community and "its" piece of earth is central to the phenomenon of nationhood itself, as Anthony D. Smith demonstrates.[7] The cultural question at the heart of nationalism—"Who are we?"—is at the same time always a geographical question—"Where are we?" or "What is our place?" Discourses of national identity seek not only to define the nation's character but also to delineate its homeland by articulating the nation's relationship to a specific territory. "The creation of nations," Smith observes, "requires a special place for the nation to inhabit, a land 'of their own.' Not any land; an historic land, a homeland, an ancestral land." Homelands are constructed by infusing physical terrain with national meaning, transforming landscape into "ethnoscape."[8] Through a wide range of practices—school curricula, the arts, heritage tourism, public spectacles—historic sites and monuments are naturalized and natural features historicized by national entrepreneurs seeking to inculcate a profound identification with an ancestral homeland. In this manner, "long barrows in Wessen, stone circles in Brittany or the Orkneys, *kreml*-fortresses in Russia, temples in Greece and Italy, *tells* in Iraq and Syria, have all entered the imaginative fabric of the community over the centuries, by appearing to fuse with a surrounding nature and become one with the habitat."[9] National identity, as Simon Schama puts it, "would lose much of its ferocious enchantment without the mystique of a particular landscape tradition: its topography mapped, elaborated, and enriched as a homeland."[10]

In England, the iconic rural landscape, with its patchwork of neatly cultivated fields, hedgerows, and bucolic villages, is seen as embodying the national virtues of order, stability, insularity, and good stewardship.[11] In Wales, early twentieth-century nationalists extolled the rugged and remote northern hill country as the defender of Welsh language and tradition against the onslaught of Anglicizing modernization.[12] Zionists planted forests in Palestine to enact a redefinition of Jewish identity away from "rootless cosmopolitanism" to an earthy connection with the land, to mark Jews as the rightful cultivators of the land who were making the desert bloom after centuries of mistreatment. The ritual of tree planting by Israelis and Jews visiting from abroad continues today, as does the promotion of nature tours aimed at teaching love of country by interweaving natural and social history.[13] In nineteenth-century America, artists, writers, and politicians seeking to define a unique identity for the upstart nation discovered that wildness was what distinguished America's nature from Europe's and that the splendor and antiquity of places like Yosemite Valley could be invoked against the Old World's indisputably superior cultural and historical heritage. Wilderness evoked America's frontier and its pioneer past, too, which was seen as having shaped the distinctively rugged but democratic and entrepreneurial character of its citizens.[14] The project to define national identity intersected with the Romantic celebration of the sublime to yield a dominant construction of nature as wilderness, valued for its pristine and untouched qualities, and institutionalized most prominently in the national park.[15]

The Power of Environmental Narratives

The natural sciences produce their own distinctive constructions of nature. Generations of scientists have articulated a series of environmental narratives or orthodoxies claiming to explain the causes and consequences of environmental change and to prescribe appropriate human relationships to nature: the narratives of deforestation, soil erosion, the population bomb, the limits to growth, global climate change, and so on.[16] Although these narratives come and go as new research upends prior certitudes, nonetheless each narrative in its heyday commands tremendous influence over perceptions of nature and agendas for nature management.

Scholars in the interdisciplinary field of political ecology have shown that those who control environmental narratives also wield power over land and natural resources. In many African colonies, colonial officials used scientific theories of soil erosion and conservation to justify strict control of native farmers' land use practices or even forcible removal from their lands.[17] In colonial and postcolonial Java, German-style "scientific forestry," with its notions of conservation and sustained yield, has underpinned state policies limiting access to valuable teak forests for millions of forest villagers.[18] Often the authoritative language (or "Big Talk") of the natural sciences[19] is reinforced by Romantic narratives of nature as pristine, Edenic wilderness. The result is a potent mixture that clashes with indigenous understandings of nature as a lived-in landscape of labor and as a reservoir of cultural, religious, and historical meanings.

Ever since the 1992 United Nations Conference on Environment and Development in Rio de Janeiro (the Earth Summit) inaugurated the present era of global environmental politics, dominant environmental narratives have been transmitted through the provision of environmental aid by rich countries to the poor. In the wake of Rio, not only conservation organizations but also mainstream development donors such as the UN and the World Bank have been doling out environmental aid dollars to developing countries and in the third world. Since the collapse of Communism, such aid has gone to the second world as well. These flows of money and expert assistance are structured by the master environmental narratives of biodiversity conservation and sustainable development.

The term *biological diversity*—natural variation at the levels of genes, species, and ecosystems—emerged as a scientific response to the shocking rates of species extinction discovered in the 1970s and has become a household phrase. Yet even as the clarion call to prevent global biodiversity loss sounds ever louder, scientists are far from agreement on how to define and measure biodiversity, let alone how best to preserve it. "Biodiversity shines with the gloss of scientific respectability," observes David Takacs, "while underneath it is kaleidoscopic and all-encompassing: we can find in it what we want, and can justify many courses of action in its name."[20] To be sure, the moral imperative to stop the annihilation of natural diversity is deeply compelling; yet all too often the narrative of biodiversity loss, much like earlier narratives of soil erosion or game conservation, underpins nature management strategies that vilify local people as environmentally destructive and sometimes literally fence them out of traditional lands.[21] Armed with the discourse of biodiversity conservation, writes James McCarthy, extralocal understandings of nature "backed up by purchasing power are transforming environments around the globe, often to the detriment or against the will of local users."[22] Biological diversity has become one of the "normalizing narratives" disseminated worldwide, along with changing flows of political and economic power.[23]

Some critics of exclusionary conservation strategies have embraced the notion of sustainable development as a possible way out of this neocolonial dilemma. Sustainability was popularized in influential international documents in the 1980s and officially endorsed by UN member states at the 1992 Earth Summit.[24] Appeals to sustainability have become de rigueur in virtually all discussions of development, from the street protests of radical environmentalists to the corridors of the World Bank. The heart of the concept is the claim that broad-based, long-term prosperity cannot be achieved without integrating ecological and social concerns into the traditional economic focus of development. The sustainability paradigm underpins the hope that biodiversity itself can be used as a resource for development, in part through the growing business of ecotourism.[25]

Thus far, however, effective implementation of such strategies has proved maddeningly elusive. Indeed, as legions of critics have noted, environmental sustainability is a profoundly ambiguous and contested notion, and there are no universally accepted blueprints for its implementation. Unlike, for example, the so-called Washington consensus on structural adjustment and privatization, there is no similar consensus—not even among donor countries and institutions, let alone among recipients—on what sustainable development should actually look like in practice. Sustainable development has been denounced as a politically correct but ultimately meaningless term, "sufficiently vague to allow conflicting parties, factions and interests to adhere to it without losing credibility."[26] Thanks to this vagueness, radical critics argue, projects in the developing world can be cloaked with the mantle of sustainability while continuing to serve the interests of wealthy countries. Rather than protecting nature and empowering local communities, the pursuit of sustainable development, along with biodiversity conservation and the other components of global environmentalism, simply extends the scope of technocratic managerial strategies to the entire planet.[27]

Contesting Nature on the Road to Europe

While most work in political ecology has focused on the global South, scholars are beginning to trace the genealogy of environmental narratives in Eastern Europe and to examine the impact of both Western aid and the ideological, technological, and structural legacies of Communism on nature management and environmental practices.[28] The lion's share of aid to the region has been directed at preventing cross-border air and water pollution, but a significant portion is targeted at biodiversity conservation and sustainable rural development. Yet even as European institutions and governments have been exporting sustainability, the daunting challenge of eastward enlargement has intensified debate over how best to reform the EU's costly and cumbersome system of agricultural price supports.

The production-based subsidies at the heart of this system have promoted agricultural intensification, which in turn has caused a loss of biodiversity. As agricultural productivity and surpluses have grown over the last twenty-five years or so, Europe has been in the throes of a "postproductivist" transition. Its increasingly urbanized and affluent citizenries have rallied to the cause of preserving Europe's traditional agrarian landscapes, and the countryside has been reimagined as a site no longer primarily for agricultural production, but rather for environmental, aesthetic, and recreational amenities. Most European environmental organizations have sought to capitalize on this value shift by advocating a policy of subsidizing traditional, "low-impact" farming methods. Because most of Europe's land has been shaped by centuries of human use, its current plant and animal populations largely depend on the seminatural habitats created by traditional farming, so promoting nonintensive agriculture is at the same time a strategy for protecting biodiversity.[29] But others reject this approach on both ecological and economic grounds, proposing instead the more radically postagrarian strategy of taking as much agricultural land as possible out of production altogether and replacing farming, traditional or otherwise, with management for natural processes and wilderness creation.

European debates over the contours of the postproductivist transition were borne into Latvia in the 1990s on the wings of assistance for nature management and rural development. By examining Latvian responses to several Western-funded aid projects, this book focuses on local perspectives and especially the diversity of these perspectives, as political ecologists have done for many third world areas.[30] During the 1990s, some Latvians embraced the radical postagrarian agenda, but many others interpreted such strategies as attempts implicitly or explicitly to exclude Latvians from their ethnoscape. Both the resistance to and the appropriation of global environmental narratives were shaped by historically rooted discourses of national identity and developmental destiny. Debates among Latvians over how to conserve biodiversity and promote rural development were fundamentally debates about who Latvians are and what their proper relationship to nature should be.

Dueling Discourses of Nation and Homeland

Since the first articulations of Latvian nationhood during the National Awakening of the 1850s, Latvians have told two kinds of stories about themselves and their homeland: an outward-looking one and an inward-looking one. The founding fathers of Latvian nationhood sought liberation from postfeudal penury and the definition of the Latvian as a peasant. Some imagined Latvians as cosmopolitan middlemen whose fortunes lay in exploiting their homeland's littoral geography of transit. This discourse of *liberal internationalism* constructed the homeland as a crossroads or bridge between East and West.

Internationalism was the prevailing orientation in Latvian cultural and political life during the fin de siècle. But after defending an unstable independence from Germany and the Soviet Union in 1918, Latvia, like the other new postimperial nation-states established by the formerly dominated "small peoples" of Eastern Europe, was swept by a wave of populist peasantism. The new bourgeois ruling elites saw a landed peasantry as the best bulwark against the threat of Soviet Bolshevism as well as the domestic socialism that had dominated Latvian political leanings since the turn of the century. At the core of their nation- and state-building project, Latvia's founders enacted a radical agrarian reform designed to distribute small parcels of land to as many Latvians as possible with the explicit aim of creating a "nation of farmers." Land reform was accompanied by a discourse that linked the national character to peasant values and laboring on the land. The hardworking family farmer as the quintessential Latvian citizen and the dispersed solitary homestead as a quintessential feature of the Latvian landscape were celebrated in the arts, in museums and monuments, and in public spectacles. This inward-looking *agrarian nationalist* discourse identified closeness to nature as a central element of Latvianness and in particular the closeness obtained through agrarian labor.

Under Soviet rule, the "nation of farmers" and its agrarian ethnoscape came under attack by the theoretical forces of Marxism-Leninism, with its frank hostility to all things rural, and the material forces of agricultural collectivization and modernization. Paradoxically, however, the agrarian nationalist discourse continued to be reproduced both through official practices—by the promotion of folklore and ethnic heritage, by subsidies for a very large agricultural sector, and by semiofficial tolerance of private plot farming—and through the soft dissidentism of a crypto-nationalist landscape protection movement. Indeed, throughout the Soviet Union, despite the antinationalist policies implemented by Soviet leaders from Lenin to Gorbachev, many factors helped to maintain or even produce national consciousness among national minorities. The institutions of Soviet ethnofederalism reified nations as administrative entities endowed with their own territory, and affirmative action policies promoted titular elites in the administrative structures of these units. Personalistic networks flourished in response to the endemic shortages that plagued state-socialist economies. The essentializing, dichotomizing, and totalizing impulses of Communist ideology fostered an "us versus them" worldview.[31] But as I show in chapter 2, it was through local discourses and practices like the Latvian Great Tree Liberation Movement of the 1970s that particular narratives of identity and homeland were reproduced in the face of Soviet repression.

In the 1980s, the Latvian independence movement was ignited by a mass movement to block construction of a dam on Latvia's "river of destiny" (the Daugava), and the language of this movement echoed the 1970s discourse of defending the ethnoscape, which in turn reproduced the tropes of the 1860s and the 1920s. To this day, Latvians claim a special closeness to nature as a key defining element of national identity, and the nation of farmers remains the dominant image of the Latvian nation. According to a Latvian sociologist, "It would be difficult to find another self-reference within Latvian culture that is as capacious and enduring" as that of the peasant culture or nation of farmers.[32] And the dominant construction of homeland and of nature remains the agrarian ethnoscape: a cultivated landscape of labor that mutually constitutes and is constituted by the Latvian national character and serves as a reservoir of national history and ethnographic uniqueness. This agrarian hegemony is not unique to Latvia; indeed, throughout post-Communist Eastern Europe there is widespread adherence to what Ernest Gellner wryly dubs the "potato principle," or the notion that "agrarian populations are morally more significant than industrial populations."[33]

Agrarian nationalists envision Latvia's post-Soviet development by looking backward to the peasantist golden age of the 1930s. Agrarianism is seen as a central component of what Daina Stukuls Eglitis calls the discourse of temporal normalization, which embraces "a conservative restorationist return to the institutions of the independent interwar state." But in the post-Soviet era, this discourse has been seriously challenged by a rival discourse of spatial normalization, in Stukuls Eglitis's terms, which envisions a "return to Europe" and "the construction of a liberally oriented, modern (read Western) European state."[34] In Latvia, as throughout the region, the cosmopolitan narrative of the "bridge between East and West" has enjoyed a renaissance since the fall of Communism. Given the power of the neoliberal economic orthodoxies disseminated by international aid donors, and given the inescapable presence of an integrationist European Union on one side and an unpredictable Russia on the other, many East Europeans have sought to define their nation's place in this rapidly evolving continent by resurrecting the language of bridges and crossroads. Vaclav Havel, addressing the European Parliament in 2000, described the Czech Republic as "a country in the very center of Europe, which has for centuries been a crossroads of multiple European spiritual currents as well as of geopolitical interests."[35] Even in Kaliningrad Oblast, the grimly isolated Russian exclave and former military outpost squeezed between Lithuania and Poland, in 2003 a regional official envisioned the district's future as a "bridge for the integration of Europe and Russia."[36]

But perhaps nowhere has this trope figured more prominently than in Latvia and its historical cousin Estonia, the erstwhile Swedish, German, and Russian subjects and "most European" of Soviet republics.[37] In both countries, dominant center-right politicians have largely toed the neoliberal party line on trade and macroeconomic policy and other matters of greatest interest to international lenders. Governing elites have pitched Latvia as multicultural transit hub or Baltic Hong Kong and Estonia as a "political mediator," as a "country that identifies itself with Europe but also knows (and remembers) Russia."[38] An Estonian government proclamation declared in 1992, "We are convinced that Estonia's geopolitical location, historical ties and current political situation enable Estonia to become a bridge, or a land of contacts."[39]

To be sure, the outward-looking discourse does not always extend a welcome in all directions. Geopolitical realities—imperial domination or its lingering fallout in one direction, rosier economic prospects in another—can obviously make one end of the bridge more attractive than the other at any given historical moment. Many Latvians and Estonians have embraced the narrative of Europeanness to promote openness in the westward direction only, while defending a cultural boundary to the east. "Estonians have always identified themselves as belonging to the West, and this was recognized by other Soviet nations who regarded the Baltic countries as the 'Soviet West,'" declares Peeter Vihalemm in a volume tellingly entitled *Return to the Western World*.[40] Estonian intellectuals have adduced historical, cultural, and even genetic evidence to prove Estonia's historical participation in the broad cultural and economic currents of "Europe" and have identified this participation as marking Estonians' difference from Russians.[41] Scholars and politicians alike have eagerly endorsed Samuel P. Huntington's theory of the "clash of civilizations," which places Estonia and Latvia at the very edge (but plainly on the "right" side) of the alleged cultural borderline that for centuries has separated Western Christians from Muslim and Orthodox peoples.[42] According to Marju Lauristin, a former government minister, "For Estonians and other people with a Western mind-set, living under the Soviets meant a 'clash of civilizations' inside the mind of every single individual."[43] For Lauristin and many of her peers, Estonia's transition from Communism was unambiguously unidirectional: it was a re-Westernization, or as the popular East European rallying cry of 1989 had it, a "return to Europe."

Estonia's borderland geography, like Latvia's, has been celebrated sometimes as a true crossroads, open in all directions, and sometimes as an exclusionary cordon sanitaire. But even in its more Russophobic mode, the narrative of Europeanness is much more inclusive and welcoming than its chief discursive rival: a homeland narrative, in Merje Feldman's phrase, that sharply distinguishes a culturally pure nation from the outside world. Resisting European influences as well as Russian, this narrative claims for Estonians "a unique relationship with the Estonian territory that gives them a primordial moral right to that space.[44]

Similarly, today many Latvians—arguably, most—believe that the survival of the nation depends on keeping rural space filled with Latvian dwellers. Indeed, even more than in 1918, today the inward-looking agrarian orientation derives special potency from the post-Soviet demographic context; in 1998 ethnic Latvians made up only 56 percent of a population of just under 2.5 million, and even less in the major cities, with the remainder comprising a mixture of mostly Russian-speaking Slavs.[45] Haunted by the specters of demographic annihilation and a potential fifth column, many Latvians defend the relatively ethnically homogeneous countryside as "the last and only refuge of the Latvian people."[46] In private conversations and in the media, politicians, poets, and former collective farm workers lament the wrenching liberalization that has forced many families to abandon their farms, or at least to shrink production to tiny subsistence plots.

In Latvia as throughout the post-Communist region, the emptying of the countryside was symptomatic of land surplus: before EU accession, apart from the fertile central regions, most agricultural land had little or no market value, and much land had no owner or claimant. Unlike Western Europe, Latvia lacks a wealthy urban stratum willing and able to support farmers as guardians of the landscape. To keep people in the countryside, therefore, value must be created in an economically devalued rural space. This, then, is the fundamental challenge of post-Communist rural development. In confronting it, Latvian environmental professionals have embraced the narratives of biodiversity and sustainability, but they have most often filtered them through the prism of agrarian nationalism, which imagines farmers as not only the truest Latvians, but also the best stewards of nature. In Latvia, as throughout most of Europe, current biodiversity resources reflect centuries of human use in the form of cultivation, livestock herding, forestry, and hunting. Like the majority of their Western counterparts, most Latvian environmental professionals see the values of nature as deriving not primarily from pristine wilderness but from cultivated landscapes: from the fields, meadows, and carefully managed forests that constitute the agrarian ethnoscape. Out of the array of imported policy ideas, many Latvian conservationists have shown the greatest enthusiasm for subsidies for traditional farming and tourism based on ethnographic heritage and the pastoral landscape.

Like all hegemonic discourses, agrarian nationalism has made it hard for many Latvians to imagine any other future for Latvia than as a "nation of farmers." Among environmental and development professionals, however, an outspoken minority has sought to radically re-envision Latvian nature in terms of protecting global biodiversity, transcontinental flyways, and wilderness tourism. These internationalist-oriented conservationists—most prominently, the Latvian branch of the world's largest conservation organization, the World Wide Fund for Nature (WWF-International)—have seen Latvia's land surplus, together with Europe's postproductivist transition, as an opportunity for more radically postagrarian change. Their goal is to detach the value of land from farming and to link it instead to global norms and markets for nature as a reservoir of biodiversity and a destination for ecotourism.

Latvian responses to international environmental aid initiatives have thus been structured by the discursive divide between agrarianism and liberalism. A proposal to bring Latvia's flagship national park in line with international criteria by excising large tracts of agrarian ethnoscape and reducing the park's territory to a biological "core zone" was defeated, as was a plan to create a new national park in a territory historically open to subsistence uses by local residents. The most passionate resistance was provoked by a major project to incorporate international sustainability norms in a fundamental overhaul of forest management practices, with their deep roots in agrarian nationalism, German scientific forestry, and Soviet productivism. Agrarian nationalism has provided a potent language with which to resist the normalizing discourses of global biocentric environmental management, but liberal internationalism, in turn, has provided ammunition for a counterhegemonic challenge to agrarianism. Today's internationalists are again imagining Latvians as middlemen in the East-West trade in nature and are promoting a globalization of the ethnoscape: opening Latvia's nature and countryside—and not just the cities and seaports—to international influences. For agrarians, globalizing the ethnoscape represents a neocolonial exclusion of Latvians from their proper place in nature for the benefit of foreign interests; internationalists hope that it can become instead an act of redefinition by and for Latvians themselves.

Contesting the Nation from Within

This book, then, is about what happens to the national "soul" and homeland when a self-defined nation of farmers escapes the frying pan of Soviet domination and collectivized agriculture only to plunge headlong into the fire of an increasingly postagrarian Europe. The book can be read in at least two ways. For students of political ecology, it is an exploration of the influence of local discourses—and particularly those of national identity—on the "creation, legitimization and contestation of environmental narratives"[47] in post-Communist Latvia. For students of nations and nationalism, it examines social constructions of nature as a window onto the creation, reproduction, and contestation of national identity.

For readers in the latter camp, some clarifications are in order, given the multiple, theoretically contested, and often imprecise uses of the terms *nation* and *nationalism*. A useful definition for my purposes is the one proposed by Jeff Chinn and Robert Kaiser: a nation is "a backward-looking community whose members share a primordial sense of common genealogical and geographic origin," and it is also a "forward-looking, modern political interest group whose members share a desire to control their common destiny."[48] I consider everyone who shares these beliefs and desires to be a Latvian nationalist. Most Latvians, including most of those whose views I represent in this book, view nations as real entities with deep historical roots; they believe the cultural survival of nations to be a good thing; and they consider the nation-state a desirable political arrangement for ensuring this survival. Unlike the situation in many other post-Communist states with historically weak or contested articulations of national identity (Ukraine and Belarus, for example), the desirability of a "Latvian Latvia" is generally accepted. Agrarians and internationalists alike are committed to preserving a Latvian state in which the sole official language and dominant culture are Latvian and in which Latvians are the primary political shapers of their own destiny.

Yet while Latvians are united in the belief in common origins and the "desire to control their common destiny," they tell very different stories about those origins and propose very different visions of that destiny. I argue that the tension between these different stories—in particular, between the inward- and outward-looking stories—is not unique to Latvia, but is a vital and all too often overlooked component of the phenomenon of national identity in general.

In Jonathan Franzen's novel *The Corrections*, the erstwhile Soviet dissident, Lithuanian freedom fighter, and politician-turned-swindler Gitanas Misevičius tries to explain to his American business partner why politics no longer makes sense to him since his nation's independence from Soviet rule. "How Lithuanian we all felt," he says, "when we could point to the Soviets and say: *No, we're not like that*. But to say, *No, we are not free-market, no, we are not globalized*—this doesn't make me feel Lithuanian. This makes me feel stupid and Stone Age. So how do I be a patriot now? What *positive* thing do I stand for? What is the *positive* definition of my country?"[49] Gitanas's angst squarely targets a central feature of the post-Communist transformations. Would-be nation-states emerging from Soviet domination have faced not only the vast challenges of state building and economic reorienting, but also the challenge of building a national identity. This challenge demands "a reversal of the practice of being defined by others to shouldering the responsibility of self-definition."[50] Nationalists like the fictional Gitanas have had to ask themselves: now that we are no longer defined by our imperial oppressors and can no longer define ourselves through our opposition to that oppression, who are we? What positive thing do we stand for?

Gitanas's question highlights a vital dimension of nationalism that is ignored by many scholars of the subject. Typically, national identity is understood solely as an effort to draw and police boundaries between members of the nation and outsiders. Identity discourses are seen as tools with which groups define themselves "not by reference to their own characteristics but by exclusion, that is, by comparison to 'strangers.'"[51] Thus, most accounts of nationalism in post-Soviet Latvia or Estonia—scholarly and journalistic alike—focus almost exclusively on border-policing practices: on the tense relations between titular nationals and Russophone minorities, and on the contestation of policies regarding citizenship, language, and integration of minorities.[52] Latvia and Estonia were the only two Soviet successor states that did not grant automatic citizenship to everyone residing within their borders in 1991, largely due to the perceived demographic and political threats posed by the very high proportions of nontitular nationals living there (48 percent in Latvia and 39 percent in Estonia in 1989). The relatively exclusionary policies adopted to ensure the survival of Latvian and Estonian as the sole official languages of those countries, and to ensure the dominance of ethnic Latvians and Estonians in political institutions, have been the subject of endless scholarly studies and diplomatic negotiations (not to mention political posturing by the big neighbor to the east).

Marking difference from others (and usually from one crucial Other) and determining who does and does not belong to the imagined community are undeniably central to the phenomenon of nationhood. But national identity is not simply a reaction against otherness, much less against a particular Other. It is also about the broader and more complex problem of maintaining a distinctive sense of col-

lective self in relation to the outside world. Nations newly emerging from imperial confines—and especially from the confines of an insular geopolitical system
like the former Communist bloc—must craft strategies not only for managing
their own economic and political affairs, but also for reinventing and preserving
their national sense of self in the process. And they must do so in a context of
openness to a suddenly expanded array of outside forces. In facing these challenges, a negative demarcation of difference from the Other—"How are we Latvians (Lithuanians, Ukrainians, Kazaks) not like Russians?"—does not suffice.
Rather, nationalists must also project a substantive self-imagining: "What is the
positive definition of my country?"

No nation is a monolithic whole, of course, so nationalists inevitably imagine
their community and map their homeland in diverse ways. To understand the
phenomenon of nationalism, we must look not only at border skirmishes but also
at competition within the nation between rival narratives of place and identity.
National identity needs to be analyzed not simply as a monolithic sense of difference from others, but also as an inevitably fractured and contested understanding
of the self. This does not mean abandoning the question of boundaries and difference, but rather posing it in a more expansive, less binary way. Indeed, the most
profound discursive fault line within the nation often concerns the negotiation of
openness to the outside world.

There are essentially two strategies available to nationalists seeking to define
and preserve a distinct community while engaging with the wider world. Either
they can urge the nation to retreat inward and erect barriers against foreign con-
tamination, or they can devise ways to "sell" itself to the world. Although the latter, outward-looking strategy embraces openness, it is as much a part of the phenomenon of national identity—as much a strategy for national survival—as the
inward-looking one. Thus Franzen's Lithuanian nationalist believes it would be
"stupid and Stone Age" for his compatriots to resist the forces of marketization
and globalization as they had once resisted Sovietization. In imagining the future
face of his homeland, he looks (albeit somewhat ruefully) outside its borders.
"The best future I can hope now for my country is that someday it looks more like
a second-rate country in the West," Gitanas reflects. "More like everybody else, in
other words." "More like Denmark," suggests his American friend. As Michael
Billig puts it, national identity "involves a dialectic of inwardness and outwardness. The nation is always a nation in a world of nations. 'Internationalism' is not
the polar opposite of 'nationalism,' as if it constitutes a rival ideological consciousness. . . . An outward-looking element of internationalism is part of nationalism and has accompanied the rise of nationalism historically."[53]

In Latvia, even debates over the political and social integration of Russophones reflect this dialectic of inwardness and outwardness; they are not merely about othering. Latvians have argued fiercely among themselves about how far they should go in satisfying international demands to liberalize their exclusionary language and citizenship policies that have formed key stumbling blocks in Latvia's relations with both Russia and the West. The inward-outward divide has structured Latvian views regarding not only the Russian Other (for or against integrating ethnic minorities into the national community), but also Latvia's relationship to the international community more broadly.

During a crucial debate over citizenship in June 1994, for example, the center-right governing coalition argued for removing exclusionary naturalization quotas on the grounds that Latvia had no choice but to bow to supranational forces in one direction or another. If Latvia did not adopt the reforms desired by Western institutions, they claimed, it would be denied membership in Europe and, by default, be pulled back into Russia's sphere of influence. As an editorial in the liberal daily *Diena* put it: "In this situation we have two alternatives—democratic Europe or imperialistic Russia."[54] The Latvian president warned that by clinging to naturalization quotas, Latvia was "isolating itself from the other countries of the world and would thus remain alone with its problems." Supporters of quotas, in contrast, maintained that resisting foreign influences in both directions was a viable option. They saw the choice facing Latvia "not as Europe versus Russia, but rather as Latvia versus Russia or a Latvian Latvia versus a Russian Latvia." National survival would be ensured not by integrating Latvia into Europe, but by resisting "meddling in Latvia's international affairs" on the part of any outsiders. In signaling their opposition to Western intervention, these inward-looking nationalists went so far as to picket the U.S. Embassy in Riga.[55]

Thus even in struggles over boundaries and membership we can observe nationalism's dialectic of inwardness and outwardness. But we can gain a deeper understanding of national identity as a contested negotiation of openness to the outside world if we broaden our scope to examine struggles over the ethnic meanings that turn nature into homeland. On the one hand, writes Sidney Plotkin, "nothing is more intimate in the life of a community, or more reflective of its most sacred commitments and prerogatives, than its treatment of land."[56] This is particularly true in Latvia, where occupations in nature management, agriculture, and forestry are filled almost exclusively by ethnic Latvians.[57] Land and nature are uniquely significant in Latvia also because Latvians embrace neither a belief in religious election nor a strong sense of an ancient, glorious age of heroism—the other two critical "ethno-symbolic resource repertoires" of nationalism, according to Anthony Smith.[58] Indeed, the myth of seven hundred years of slavery under the feudal yoke is much more powerful than that of a distant golden age.[59] As Edmunds Bunkše contends, "The dominant element in Latvian culture is nature, not history; not stories of deeds and events (although these are not lacking) but evocations of being; of life in particular nature-culture settings."[60]

On the other hand, in the present era of global environmental politics, the intimate realm of land and nature has been deeply penetrated by international forces. Examining the nexus of nature, national identity, and globalization illuminates identity as a people's internally contested, substantive self-definition in relation to the outside world in ways that merely focusing on policing the nation's external borders cannot. Curiously, however, scholars within the nationalism canon have paid scant attention to constructions of nature (with the prominent exception of Anthony Smith). Political scientists and sociologists, on the one hand, have largely overlooked the links between place and nation.[61] Geographers, on the other hand, have only recently returned to the study of nations and nationalism.[62] And in doing so they have typically reduced the meaning of homeland—like that of national identity overall—to the marking of otherness: to the quest to satisfy a yearning for wholeness by "sustaining cultural boundaries and boundedness" and by protecting "exclusive, and therefore, excluding, identities against those who are seen as aliens and 'foreigners.'"[63]

I argue that homeland narratives do more than serve this negative, exclusionary function, for they also allow a national community to project a positive understanding of itself. In imagining the homeland, nationalists manipulate "collective perceptions, encoded in myths and symbols, of the ethnic 'meanings' of particular stretches of territory," to provide "'maps' of the community, its history, its destiny, and its place among the nations."[64] Thus in Latvia, both the internationalist and agrarian discourses articulate substantive notions about the Latvian national character, map the ethnic meanings of the homeland, express a continuity with a national past, and envision a developmental trajectory for Latvia's future. In each discourse, cultural imagining shapes and is shaped by geographical imagining; ideas about who Latvians are and where they are going cannot be separated from ideas about Latvians' relationship to the natural terrain of "their" territory.

My focus on contending discourses is underpinned by the premise that the *substance* of national discourses is important. Yet much of the literature tends to dismiss such substance as irrelevant, arbitrary, or mere false consciousness. According to Ernest Gellner, for example, "The cultural shreds and patches used by nationalism are often arbitrary historical inventions. Any old shred and patch would have served as well."[65] I maintain instead that the particular stories that members of a nation tell themselves—stories about who they are, where they are, and what kind of future they desire—*do* matter. Articulating a positive definition of self is essential to defining the self negatively against the Other. To convincingly mark their difference from Russians or Germans, Latvian nationalists must specify what, in particular, is deemed special about Latvians, what it is that "we" do differently from "them." Moreover, such identity discourses shape people's perceptions of their interests and visions of the future in important ways. The claim that "we are not like Russians" does not in itself promote a specific policy stance on agricultural support or trade policy. But the claim that "we are a nation of farmers" or "we are a nation of cosmopolitan middlemen" certainly does.

Scholars often emphasize the commonalities among many national discourses, the universal rhetoric of competition and equality in which nationalist claims are expressed.[66] But national discourses also vary. As Schama notes, "Not all cultures embrace nature and landscape myths with equal ardor. What the myths of ancient forest mean for one European national tradition may translate into something entirely different in another."[67] Many agrarian discourses represent the symbiotic links between nation and homeland as a function of kinship, of blood and soil: *land* is transformed into *homeland* through the bodies and blood of the ancestors. The trope of blood is sometimes intertwined with the trope of labor. In conflicts over agricultural decollectivization in post-Communist Romania, Katherine Verdery argues, a kinship-based sense of Romanian identity mutually reinforces a celebration of personal labor: "Land, ancestors, work, and the sense of self are interwoven."[68] The Latvian case differs from this common pattern in that constructions of nature by agrarian nationalists invoke the symbolism of labor, but not that of blood. Notions of rootedness in place are far more central to Latvians' self-understanding than notions of kinship. This rootedness is seen as springing from agricultural labor and the cultivation of nature rather than from the soil as repository of ancestors' bones. Connectedness to one's ancestors is ensured by preserving and reproducing the landscapes shaped by their labor.

The particular stories Latvians tell about themselves matter because of the crucial role of discourse—the durable ideas about nation and homeland that are reproduced from generation to generation, the tropes through which these ideas are voiced, and the practices in which they are embedded. National discourses cannot be dismissed, as instrumentalist theories of nationalism would have it, as mere false consciousness, tools of elite manipulation, or functions of economic or institutional factors. Discourses help to define interests, rather than merely reflecting them. As Rogers Brubaker observes, citing Max Weber, "Very frequently the 'world images' that have been created by 'ideas' have, like switchmen, determined the tracks along which action has been pushed by the dynamic of inter-est."[69] Indeed, discourses constitute perceptions of reality by making it hard to think outside their boundaries. Within a given discursive space, some strategies seem reasonable and others absurd or impossible.[70]

Like Brubaker, I seek to demonstrate "how particular cultural idioms—ways of thinking about talking about nationhood . . . were reinforced and activated in specific historical and institutional settings; and how, once reinforced and acti-vated, these cultural idioms framed and shaped judgments of what was politically imperative."[71] I focus primarily on the intellectuals, politicians, and other leaders who do most of the activating, reinforcing, framing, and shaping of discourse. Drawing upon published primary texts, interviews, and ethnographic observa-tion, I show how the dueling idioms of agrarianism and internationalism have provided the "tracks" on which interests have been defined and defended in suc-cessive periods of tremendous uncertainty: the National Awakening of the mid-1800s, building the independent Latvian nation-state in the 1920s, restoration of statehood in 1990, and the subsequent "return to Europe."

While I share the constructivist belief that ideas shape interests, I do not em-brace the corollary notion that national discourses are constantly being "in-vented," or fabricated anew, at any given historical moment.[72] Constructivist the-ories see national identity as radically fluid and shifting, continually manipulated by individuals engaging in a "constant process of defining and redefining them-selves."[73] At the other extreme are theories, variously referred to as perennialist, primordialist, or ethnosymbolist, that view identity discourses as rooted in prena-tional ethnic myths and beliefs that remain relatively stable over very long periods of time.[74] In this view, history is more "full" than "empty," as Smith puts it. When cultural entrepreneurs manipulate discourse, they are operating not on a tabula rasa but in a conceptual field already saturated with durable values and beliefs; nationalists are more like archaeologists than engineers. Although nationalists undeniably manipulate the "truth content" of their inventions or rediscoveries, Smith and others argue, these rediscoveries can take hold only when their "mem-ory content" is rooted in preexisting legends, myths, and symbols.[75] These are continually being questioned and reinterpreted, "but all this questioning and re-placement is carried on within definite emotional and intellectual confines," and the images pieced together and disseminated by each generation "become the of-ten unconscious assumptions of later generations in whose social consciousness they form a kind of rich sediment."[76]

My reading of Latvian identity discourses comes down on the side of durability against malleability, at least in the modern period. While I bracket the question of the relationship between national discourses and premodern myths, I do find remarkable continuity in Latvian ways of talking about nationhood throughout the one hundred fifty years since the National Awakening, at times in the face of considerable forces for change: the Soviet assault upon agrarian identity, for example, through agricultural collectivization and liquidation of homesteads. The tropes of the 1860s resurface essentially verbatim in the 1930s, the 1970s, and the 1990s; they are not fabricated de novo by each generation, although their specific meanings vary with the changing context. These inherited ways of talking about the nation and the homeland delimit the conceptual universe within which Latvians typically think about themselves and their land; they provide the palette of colors with which Latvians attempt to paint their vision of the future.[77]

While emphasizing the longevity of identity conceptions, however, one must reiterate that these discourses are not monolithic. The universe of myths, symbols, and tropes from which nationalists can convincingly draw is limited by their cultural inheritance, but individuals can select different elements from within that universe and interpret those elements differently. Nationalism, as Mark Beissinger puts it, "is not simply about imagined communities; it is much more fundamentally about a struggle for control over defining communities, and in particular, for control over the imagination about community."[78] Not only has the content of the Latvian agrarian and internationalist narratives been reproduced across generations, but so too has the tension between them. The history of Latvian nationhood is a history of competition over nationhood.

The Plan of the Book

This study is underpinned by the belief that close investigation of local events and meanings is essential for understanding global processes.[79] The primary data were gathered during a year of ethnographic fieldwork in 1998–1999, building upon interviews I conducted in 1996 on the politics of sustainable forestry as an independent consultant for the World Wide Fund for Nature Latvia Program (WWF-Latvia). All told, I conducted approximately 160 unstructured, in-depth interviews with 135 individuals, including environmental advocates; scientists; central and regional government officials in the environmental, agricultural, and forestry sectors; local authorities and activists in rural townships; private landowners; logging executives; ecotourism promoters and providers; and international consultants. The interviews grew out of my participant observation in a range of settings, including numerous national, regional, and local-level meetings and seminars associated with Western assistance projects. The most extensive participant observation occurred at the offices of WWF-Latvia, where I spent much time visiting informally, as well as sitting in on formal strategy sessions and offering my services in drafting reports and proposals.

In interviews and other settings, I found my subjects to be exceptionally open and eager to engage in discussions, often because they were intrigued by my dual status as both a Westerner and a Latvian-American with native fluency in the Latvian language. Often, too, unlike the many jaded Rigans who have been besieged by endless streams of researchers and consultants in more glamorous realms such as minority policy or economic restructuring, residents of remote rural villages and the "patriots" of the forestry, conservation, and rural development sectors were passionately eager to discuss problems that typically receive less attention both domestically and internationally. I supplemented ethnographic data with a broad sample of primary texts. These included the writings of seminal figures from the National Awakening, canonical literary works, and other books, essays, and newspaper articles from the 1920s to the present. (All translations from the Latvian are my own.)

Part 1 traces the construction, reproduction, and contestation of the agrarian and internationalist discourses of nation, homeland, and nature from the 1850s through the 1990s. Chapter 1 situates the Latvian National Awakening in the context of the nineteenth-century national revivals among the "small peoples" of Eastern Europe, showing how Latvians made use of the two dominant tropes of this period: the "bridge between East and West" and the "peasant nation." One of the central figures of the Awakening propagated the notion of Latvians as cosmopolitan seafarers and middlemen; another sought to rekindle a lost connection with nature and landscape by appropriating the German practice of homeland studies. After independence in 1918, the new state's peasantist leaders enacted a radical agrarian reform to create a literal nation of farmers. Political and cultural figures alike imagined nature as an agrarian ethnoscape reflecting a primordial Latvian closeness to nature produced by agrarian cultivation.

Chapter 2 considers the period of Soviet rule, from annexation in 1940 through the immediate postindependence period of the early 1990s. Here I examine the Soviet assault upon the nation of farmers and its ethnoscape through agricultural collectivization, land reclamation, and homestead liquidation. I then consider Latvian resistance to this assault, especially through the Great Tree Liberation Movement launched in the 1970s by Latvia's poet laureate, Imants Ziedonis. The discourse of homeland studies is traced from the National Awakening and the interwar peasantist era through the crypto-nationalist dissent of Great Tree Liberation, the eco-nationalist protests of the glasnost era and into the first decade of restored independence. The tension between the now-hegemonic agrarian discourse and a resurrected narrative of bridges and crossroads in the 1990s is considered in chapter 3. With the fall of Communism came not only agricultural decollectivization but also a dramatic collapse in production, and the resulting agrarian-internationalist tension structured impassioned debates over land reform and agricultural supports.

Setting the stage for the three detailed case studies presented in part 2, chapter 4 explores competing Latvian responses to the new post-Soviet context of nature management and rural development. While Latvians were negotiating their accession to the EU, the latter was itself undergoing a postproductivist transition that cast doubt on the future of rural policies. The ambivalence and indeterminacy within Western Europe's "export" of the new sustainable rural development paradigm enabled agrarians and internationalists to respond quite differently to Western environmental aid initiatives.

Chapter 5 addresses the politics of national parks. Latvia's flagship national park, Gauja, was established in 1973 as a veiled expression of Latvian nationalism. When a Danish-supported project sought to enhance protection of the park's biologically rich core zone by excising large tracts of cultivated land from its territory, agrarian nationalists fought back, ultimately winning their struggle to preserve the park as a reservoir of not only biodiversity but also Latvian agrarian history and identity. And in northern Kurzeme on the Baltic Sea coast, a Finnish-supported project sought to establish a new national park. As at Gauja, park boosters invoked global forces—international standards for protected areas, access to international funding, and ecotourism markets—to promote a postagrarian developmental vision. But local residents rejected the proposal as a threat to established uses of the ethnoscape through subsistence farming, fishing, and forestry.

Chapter 6 considers a project in sustainable rural development and ecosystem restoration implemented by WWF-Latvia at Lake Pape, a remote, impoverished rural area that is also an ecologically rich wetland and transcontinental migratory corridor on the Baltic coast. At Lake Pape, WWF-Latvia promoted the radically postagrarian creation of a preagricultural landscape, or "European wilderness," by introducing wild horses and management for natural processes. But agrarian-minded critics and many area residents resisted this globalization of the ethnoscape, hoping instead to promote traditional farming and to market the area as a living museum of Latvia's ethnographic heritage.

Finally, chapter 7 examines an aggressive campaign by WWF-Latvia to introduce Western-style sustainable forestry practices that were both market-oriented and concerned with biodiversity. A majority of forestry officials and scientists passionately resisted WWF's efforts, denouncing contemporary forest ecology and neoliberal market norms as violations of Latvian traditions and a cloak for foreign interests. These critics defended an inward-turning and backward-looking conception of forest science and forest stewardship rooted in a narrative of the forest as cultivated, productive landscape and the state forester as a patriotic defender of the nation's wealth. In this case, however, WWF was highly successful in remaking Latvian nature management along Western lines.

Part 1
Imagining Nation, Nature, and Homeland

Chapter 1

By Land or by Sea

Internationalism imagines the Latvian homeland as a multicultural crossroads; agrarianism, as a primordial farmscape. Both of these tropes have deep roots throughout Eastern Europe. Radical openness to outside forces has always been an abiding reality throughout Europe's borderlands—that swath of much-contested territory stretching from the Balkans to the Baltics, "crushed between the civilizations of Europe and those of Asia," as a contemporary travel writer puts it.[1] The colonized nationalities of this region did not enjoy self-rule until after World War I, when the triumphant doctrine of national self-determination endowed them (however briefly) with statehood. But efforts at national self-definition had begun more than a century earlier, spurred by the spread of Romanticism. In Latvia as throughout the region, national revival began with an "ethnographic boom" that embraced the authenticity and vitality of peasant culture. At the same time, an outward-looking trope of Eastern Europe as a "bridge between East and West" began to be articulated by both orientalizing Westerners and East European nationalists themselves. With the rise of populist peasantism in the postwar era, the celebration of the folk idiom and of simple peasant lifeways took on an explicitly xenophobic, inward-looking political orientation, as "peasant nations" were reified in the new nation-states of the 1920s.

This chapter explores the constructions of national identity and homeland engendered in Latvia by these two dominant discourses of national revival. From the National Awakening of the 1850s through the 1930s authoritarian agrarian regime of Kārlis Ulmanis, how did Latvian nationalists seek to articulate a definitive relationship between Latvians and their natural environment?

The National Awakening: Peasants into Latvians

The early Baltic peoples—a branch of the Indo-Europeans—began to settle in the eastern Baltic littoral around 1000 B.C., although archeological evidence places the arrival of the first "proto-Baltic" settlers as early as 9000 B.C.[2] German crusaders of the Holy Roman Empire launched their first mission into the Baltic territories in the early twelfth century, began building the fortified city of Riga in 1201, and vanquished the last Baltic forces in 1290. The Livonian Confederation, a decentralized medieval state comprising the church, the crusading orders, and the urban burgher class, dominated the eastern Baltic territory until the second half of the sixteenth century, when Livonia was broken up into three separate regions under Swedish and Polish-Lithuanian rule.[3]

The eastern Baltic was a quintessential borderland, ravaged through the centuries by countless wars of expansion, colonized by multiple foreign powers, and, from the eighteenth century on, subjected simultaneously to Russian imperial rule and German cultural domination. This borderland character was most powerfully manifested in the Hanseatic trading city of Riga, founded at the mouth of the river Daugava (Russian, Western Dvina; German, Düna), which flows from the Valdai Hills of central Russia to the year-round ports of the Gulf of Riga. Situated "astride a major artery of east-west trade . . . in timber, herring, furs and hides," Riga was a city shaped by flows of goods and people, "commonly described as the most 'European' of Imperial Russia's great cities."[4]

The Russian Empire won control of the Livonian territories during the eighteenth century, but the tsars permitted Baltic German elites to retain control of local government, as well as cultural and economic life, in the Baltikum—the provinces of Kurland, Livland, and Estland.[5] The Lutheran Reformation spurred production of the first known printed material in the Latvian language, as German clerics began to translate catechisms and other religious writings into the vernacular. With the eighteenth century came the spread of egalitarian pietism among Latvian peasants by German Lutheran missionaries, increasing peasant unrest, attacks on serfdom by enlightened members of the gentry, and challenges by prosperous Latvians in Riga to the exclusionary politics of German burghers. Serfs were emancipated in the Baltic provinces in 1816, and further reforms in the 1860s allowed peasants to purchase land and established popularly elected local governance in the countryside.

Europe in the later nineteenth century witnessed what Eric Hobsbawm calls a transformation from the nationalism of existing states, or imperialist chauvinism, to that of nations seeking states, or "small-people" nationalism.[6] Miroslav Hroch defines "small peoples" as oppressed nationalities ruled by members of a foreign nationality, with no history as independent political units and lacking "a continuous tradition of cultural production in a literary language of their own."[7] With socioeconomic modernization, growing numbers of these oppressed groups became educated and upwardly mobile and began to chafe at their cultural and political subordination. This growing assertiveness was fueled by the writings of Johann Gottfried von Herder (1744–1803), a German pre-Romantic philosopher who advanced the radical definition of the nation (das Volk) as defined by language, rather than political or economic dominance. Herder himself lived in Riga in the 1760s, and it was his exposure to Latvian folk songs, in part, that led him to argue for the scientific and literary value of the languages and folklore of "primitive" peoples.[8] By the mid-nineteenth century, national revivals were in full swing throughout the region, as indigenous intellectuals from Bulgaria to Estonia began to conceive of their peoples as fully fledged nations on a par with the great imperial nations.

In the Baltikum, German elites largely viewed Latvians and Estonians not as members of nations in their own right, but rather as a cultureless, undifferentiated mass of "non-Germans" defined only as belonging to the socioeconomic stratum of the peasantry. Influenced by Herder's cultural populism, a Baltic German, Garlieb Merkel (1769–1850), was the first to describe the Latvians as a Volk in his 1796 critique of serfdom, The Latvians, Especially in Livland, at the End of the Philosophical Era. Until the mid-nineteenth century, however, publications in the Latvian language remained the nearly exclusive work of Germans, and upwardly mobile Latvians and Estonians adopted the German language and culture to rise above peasant status. Before the founding of Latvia's first university in 1919, Latvians pursued education abroad at the universities of Dorpat (Tartu), St. Petersburg, Moscow, and Berlin. Not until the 1850s did Latvian activists begin to produce a flood of Latvian-language books, pamphlets, and newspapers. Rejecting or "converting" from cultural Germanization, these "New Latvians" shockingly identified themselves publicly as Latvians and advanced Herder's notion of Latvians as a Volk (Latvian, tauta): a people defined by language, not by social class. A central project of this National Awakening was language renovation, or promotion of Latvian as a language of high culture.[9] The New Latvians, in other words, were vigorous participants in what Benedict Anderson calls Europe's "golden age of vernacularizing lexicographers, grammarians, philologists, and litterateurs."[10] At the same time, more practically oriented activists strove to confound the notion of Latvians as a peasant people by promoting their entrance into the bourgeois professions.

The National Awakening in its early years was not a mass movement, but comprised only some fifty activists without a coherent unifying ideology.[11] However, they enjoyed a wide readership, thanks to the rapid growth of literacy in the Baltic provinces. Along with literacy, modernization wrought other profound changes during the second half of the nineteenth century that helped transform the Awakening into a mass phenomenon. Spurred by the agrarian reforms and the construction of railroads linking the Baltic port cities with the Russian hinterland, industrialization and commerce took off, drawing waves of peasants to the cities. The influx of Latvian and Estonian speakers into urban centers, together with their increasing penetration of the middle-class professions, decisively tipped the scales against their assimilation into German language and culture.[12] By the mid-1880s, writes Andrejs Plakans, "a Latvian presence—an 'awakened nation'—had become a force to be reckoned with."[13]

The National Awakening meant transcending peasant status, both conceptually, by undermining the equation whereby Latvian meant peasant, and practically, by modernizing Latvian society. And yet in Latvia, as throughout the region, many National Awakeners, following Herder's lead, turned to peasant customs and folklore in seeking to construct a national history and to define the nation's particular essence or character.[14] Not only did peasants constitute the vast majority of the East European populations, but also intellectuals had learned from Herder and from Rousseau to look to the countryside for cultural authenticity and vigor as against the decadence and corruption of the cosmopolitan city.

Thus it was in nineteenth-century Eastern Europe that the discipline of ethnography was born. Nation builders fanned out into the provinces, studying peasant customs and vernacular architecture and gathering folk songs, fables, and poems. Folklore institutes were established and collections of peasant artifacts amassed in newly founded ethnographic museums. Writers and painters drew inspiration from folk themes and sought to depict the lives of peasants. In Hungary, an entire "ethnographic village" was constructed for the millennial exposition of 1896. By the early decades of the twentieth century, the Budapest bourgeoisie had taken to wearing peasant folk costumes to balls and fêtes, as had delegates to congresses of the Social Democratic Party. Zoltán Kodály and Béla Bartók published collections of Hungarian folk songs gathered in expeditions to the countryside and astonished audiences by incorporating "authentic" folk themes into their compositions.[15] Underpinning this ethnographic boom was the notion of the peasantry as "the true bearer of national culture,"[16] the "source of the continuity . . . and purity of the national spirit."[17] Thanks to their isolation in remote villages, peasants were seen as having preserved the "pure" national culture and character, unadulterated by cosmopolitan influences, thereby sparing the nation from annihilation under the foreign yoke. Folk symbols were employed to delineate the nation's cultural boundaries, to define true Polishness, Czechness, or Romanianness against the hegemonic German, Austrian, or Turkish culture.

Yet these ostensibly inward-looking celebrations of folk purity and uniqueness were at the same time manifestations of the outward gaze of nationalism, for the ethnographic boom was an international phenomenon par excellence. Artists like Bartók and Kodály were no navel-gazing xenophobes, but rather European and indeed world citizens participating in broad Western cultural trends. Europe's colonized borderlands were thoroughly hybrid zones, penetrated by international cultural currents. The folk idiom was embraced not to put up protective walls around an existing national community, but rather to mold such a community out of the ever-shifting and fluid cultural mix of empire. This embrace did assume a strongly antiforeign, defensive character in the very different cultural milieu of the 1920s and 1930s, after the peasant nations had been reified through political statehood. Even in this early phase, some national revivalists were xenophobic and defensive, but others were driven by the outward-looking desire to participate in international cultural flows and to connect the nation to larger cultural groupings. Hungarian ethnographers mined folk songs for links to the Italian Renaissance and to French ballads.[18] Polish intellectuals "wanted creative efforts to bear the mark of Polishness, but at the same time to gain international renown, make the name of Poland famous, and testify to the universal values of her culture."[19]

Cosmopolitan narratives of nationalism were often but not always Western-looking. In this region "crushed" between Europe and Asia, the tension between Western and Eastern orientations was often a crucial flash point of debate. Pan-Slavists envisioned not a European but a Slavic universality and sought to unearth "Slavic" qualities in their national heritage. In Romania, nationalist discourse was fractured into three camps: indigenists celebrated "qualities they thought peculiar to Romanians and wished to protect from the corrupting effects of imported civilizations," whereas Westernizers linked the nation historically and religiously with Rome, and pro-orientalists linked it with Byzantium and ancient Thrace.[20] Both of these outward-looking stances did seek to erect a boundary in the other direction, to fend off colonial domination by non-Europeans or non-Slavs. Yet they were not simply about exclusion and boundaries, but also about engagement with cultural forces outside the newly imagined community.

Indeed, the small nations of Europe's borderlands could not escape what Jerzy Jedlicki calls the "uneasy relationship between the national and the universal." Each national culture had to "continually test whether it is perhaps not native enough or not original enough or too imitative and secondary or, to the contrary, too inbred, ego-centric, and particular, and not European enough." In the national revivals of Eastern Europe, "cultural nationalism and cosmopolitanism thus were twins."[21] Inward-looking narratives of peasanthood were expressions of cosmopolitanism, just as cosmopolitan narratives were expressions of nationalism.

The cultural anxiety over uniqueness and universality plagued not only the region's small peoples, but also imperial Russia, eternally tormented by the question of whether or not it belonged to Europe, culturally or even geographically. The problem of Russia's "backwardness" vis-à-vis Europe provoked one of the chief schisms in prerevolutionary Russian thought, most famously embodied in the divide between Slavophiles and Westernizers.[22] Slavophiles and other Romantic nationalists inverted the problem of backwardness, messianically casting Russia as spiritually superior to a tired, decadent Europe and celebrating the peasant masses as bearers of the national essence, a bulwark against cultural Europeanization and agents of spiritual redemption.[23] Many in this camp believed that Russia's lot could be improved only through radical isolation from Europe. Westernizers, on the other hand, sought to reconcile universalism and nationalism and saw progress only through the adoption of European political, social, and economic norms and models.

The very notion of Eastern Europe, as Larry Wolff demonstrates, was invented by West Europeans perplexed by the region's ambiguous character: it was not fully European, in the eyes of Enlightenment-era travelers and armchair explorers, yet not quite Oriental either. Central to Western constructions of Eastern Europe, thus, was the trope of the bridge between Occident and Orient. "The inhabitants of the Ukraine, Russia, the plains of the Danube, in short, the Slav peoples," mused Balzac, "are a link between Europe and Asia, between civilization and barbarism."[24] This trope was embraced by not only by Western observers but also by many East Europeans themselves, such as the Russian geographer and nationalist Peter Semenov, who declared: "Chosen by God as an intermediary between East and West . . . Russia is equally related to Europe and Asia and belongs equally to both parts of the world."[25]

Thus the peasant nation and the bridge between East and West were the two crucial narratives of national revival in Eastern Europe, the two faces of nationalism's "dialectic of inwardness and outwardness." Each of these narratives yielded distinct cultural imaginings of the national character and geographical imaginings of the homeland. In Latvia these imaginings can best be illustrated by considering two of the central figures of the Awakening, Atis Kronvalds and Krišjānis Valdemārs.

Atis Kronvalds: Awakening the Nation through Homeland Studies

The Berlin-educated Atis Kronvalds (1837–1875) is revered as a cultural activist and renovator of the language, but Kronvalds himself maintained that knowledge of the homeland's natural geography was even more important than history and language in cultivating national consciousness. His geographical inclination was shared by other key figures among the New Latvians, including Krišjānis Valdemārs (to be discussed later); Juris Alunāns, author of the Latvian epic poem *The Bear-Slayer;* and Krišjānis Barons, whose monumental collection of Latvian folk songs (*dainas*) earned him the enduring title "father of the *dainas.*" In 1859 Barons published *A Description of Our Fatherland,* which was called the "first serious physical geography of the three Baltic provinces."[26]

An exploration of the New Latvian interest in geography must begin in Germany, where geographers and conservationists were beginning to theorize the links between nature and nation. German conservationist doctrines and practices were enormously influential throughout the Russian Empire, and the Baltic provinces were no exception.[27] The movement emerged during the last decades of the nineteenth century, largely in response to rapid urbanization and industrialization. Among its diverse philosophical underpinnings were environmental determinism and its counterpart, *völkisch* nationalism: the belief that nature shapes the character of the nation—*das Volk*—as much as people shape nature. Herder was an early exponent of this notion, arguing for the existence of unique ethnocultural regions that corresponded to unique national mentalities.[28] The idea was developed more fully several generations later by Germany's pioneering geographers, who retained Herder's understanding of nature as human-shaped landscape, rather than primordial environment.[29] This local or regional cultural landscape, seen as both shaping and reflecting the character of its human population, came to be known as *Heimat.* Around the turn of the century, a new *Heimatschutz* (homeland protection) movement emerged, aimed at protecting not only natural objects but also folk culture and traditional architecture. Through landscape preservation and "homeland studies," Heimat defenders sought to promote a unified national homeland consciousness as a bulwark against regional particularisms, urban degeneracy, and the incursions of foreigners.[30] By the late 1920s, the once diverse conservationist movement came to be dominated by the *völkisch* Heimat orientation. While Weimar-era conservationists did seek protection of some strict nature reserves as sites for biological field research, their sovereign aim was to protect not nature in its wild or pristine state, but rather nature as the wellspring of the German "national soul."[31]

At the same time that German conservationists were promoting homeland studies as a reaction against modernization and the erosion of rural landscapes and traditional social values, nationalists in Finland, Scandinavia, and the Baltic provinces were introducing the same subject into university curricula in the hope that "scientific knowledge and understanding of home areas could foster a sense of local and national identity." In fact, according to Anne Buttimer, "one might claim that the most intimate links between geography and regional identity were forged in the peripheral lands of Eurasia from the Balkans to Catalonia, the Celtic fringe to Karelia."[32]

In 1867 Atis Kronvalds published the first work for a Latvian audience—although in German—on homeland studies.[33] This was a short manual for teaching local and regional geography, a practice Kronvalds himself honed as a provincial schoolteacher, but its deeper purpose was national consciousness raising. The "first source [of love of the fatherland] is knowledge of the fatherland," he declared, and for Kronvalds, "knowing the fatherland" meant knowing the geography of both its natural wonders and its people's history.[34] Thus he urged in an 1886 essay:

> The young generation must be presented with living signs from our fatherland's most beautiful districts. They must be familiarized with those places that were especially sacred to our forefathers, with the hills atop which the fortresses of our fathers' fathers stood, with the rivers made red by currents of blood as the nation fought for its most sacred things, with the wondrous sea whose waves have carried Latvian ships. The young generation must be taught to revere our natural delights, to honor all manner of witnesses to antiquity . . . then the young generation will begin passionately to embrace their fatherland. Then they will no longer be strangers in the very heart of their own fatherland.[35]

Like the German Heimat enthusiasts, Kronvalds conceived of nature as intertwined with culture and history. But if the Germans invoked this linkage in order to defend the character of an acknowledged nation from the onslaught of modernization, then Kronvalds did so to tease a new national consciousness out of the gray mass of peasanthood. Before a Latvian character could be defended, it first had to be reclaimed from the obscurity of the prefeudal past, for the feudal yoke had alienated Latvians from their sense of place and of themselves; it had made them, in Kronvalds's eloquent phrase, "strangers in the very heart of their own fatherland."

Homeland studies was thus an integral element of Kronvalds's "uncompromising battle against Germanization."[36] As a former pupil recalled, just as he fought cultural assimilation by generating thousands of new Latvian words, he "knew the Latvian name for every local plant and flower."[37] Noting that the "Greeks, Romans, Germans, and Slavs" had preserved their knowledge of trees and forests in songs and poems, Kronvalds asked:

> Do Latvians alone have no remnants about the forest, about oaks, pines, and birches? Is the understanding of the forest's strange life entirely extinguished in us, do we alone no longer feel sacred presentiments in the shade of the forest? Or do we per-

haps no longer hear the rustling of the dense canopy and the cheerful singing of the birds? Do the gaily flowing forest streams no longer refresh our hearts and soul? Or were our sensation, our thoughts killed, too, when our gods were driven from the forest—so that we are no longer capable of honoring the memories of our people's youth?[38]

Closeness to nature, Kronvalds posited, was a primordial element of the Latvian national character that had been suppressed by feudalism: "In the days when our nation had not yet been starved by the burdens of slavery, was not yet exhausted in spirit and mind, our nation beautifully translated the signs of autumn . . . in sacred meditation."[39] Kronvalds enjoined Latvians to rekindle this lost knowledge by discovering the ancient "translations" preserved in folklore, and particularly in folk songs, but also by opening themselves to nature itself, which could directly transmit knowledge of the Latvian past: "If you, dear reader, should happen in some moment of leisure to be wandering in the forest at sunset, then listen closely, with the doors to your heart open wide. . . . When you have listened closely to the forest and when the Forest Mother . . . has revealed to you some secret about our people's days of youth, then do not hide it, but pass it on."[40] Nature was a "witness to antiquity," a living link to Latvians' lost past. To discover nature through homeland studies was to discover Latvianness. Kronvalds, in short, appropriated the German practice of homeland studies in order to undermine German cultural hegemony in Latvia and propagate an emerging Latvian national consciousness.

Krišjānis Valdemārs: Middlemen and Seafarers

Another giant among the New Latvians was Krišjānis Valdemārs (1825–1891), the "chief ideologist, greatest social and political activist, most notable publicist and polemicist of the New Latvians."[41] As a student of economics, Valdemārs secured his place in the hagiography of Latvian nationalism by pinning his visiting card, which identifyied him as a "Latvian," to the door of his flat in Dorpat and by helping to found a seminal Latvian-language nationalist newspaper in St. Petersburg. Valdemārs was the first writer to extend to Latvians the Herderian exhortation to pursue the scientific study of their own folklore.

His nationalist credentials are impeccable, but Valdemārs was also an ardent cosmopolitan and "an urbane man, moving easily between the centers of the empire and the Baltic Provinces."[42] While some New Latvians allied themselves with the Baltic German landed elite and while others were already beginning to advocate an independent nation-state, Valdemārs was among those who emphasized "the close cultural, linguistic and historic links with the Russian nation" and envisioned Latvia's future firmly within the Russian Empire.[43] A resident of St. Petersburg and Moscow for most of his adult life, Valdemārs was an ally of Russia's Slavophiles and a writer for the patriotic Russian newspaper *Moscow News* (*Moskovskie vedomosti*). He considered knowledge of the Russian language indispensable for Latvians' economic betterment, as it opened doors to careers in commerce and in the empire's military and civil services. Valdemārs's cosmopolitanism was correspondingly anti-German: "Knowing German but not Russian, he observed sarcastically, would help the Latvian peasant laborer to advance to the position of servant, cook, or parish clerk. With Russian, he could fully exercise his rights as a citizen of the Russian state."[44] It was the legacy of German feudalism that was making "the Latvian and the Estonian a stranger in his own fatherland," Valdemārs argued.[45]

Thus Valdemārs saw agrarian reform as a necessary step toward political emancipation and defended small landownership as "freedom's buttress."[46] It was not in agriculture that he envisioned Latvia's economic uplift, however, but rather in the exploitation of Latvia's fortuitous geographical position. In an 1886 essay entitled "Latvians' Future between the Russians and the Germans," Valdemārs enjoined the youth of Latvia:

> Learn Russian to the best of your abilities and, if possible, also German. . . . Latvia is located between great Russia and Germany, also great in terms of population. Thus it is easy . . . to learn both languages and in this manner get rich and become valuable middlemen between both of these large peoples, who in the future will need many more middlemen than up to now.[47]

Valdemārs identified as examples of successful Latvian "middlemen" members of the foreign embassy staff in St. Petersburg, as well as hotel, telegraph, postal, and railroad employees throughout Russia. His deepest enthusiasm, however, was aroused by the promise of commercial seafaring.

In an autobiographical sketch, Valdemārs dramatically recalled his first sighting of the Baltic Sea on a school outing at the age of fifteen: "The sea's majestic appearance so strongly moved the youth's spirit that for the whole three days [of the visit] he could think of nothing else but the majestic sea; plus, with childish confidence he already immediately concocted a plan or project for how to build a deeper harbor at the Roja River, so that larger boats and small ships could enter."[48] This youthful fascination blossomed into a lifelong commitment to promoting Latvian seafaring. In his first article on the topic, published in German in a Dorpat newspaper in 1857, Valdemārs correctly predicted: "A huge railway network will soon deliver manufactured goods to our seacoast from the entire wide, blossoming country, to transport them across the foaming sea to all of Western Europe and growing North America."[49] Latvia's geography represented a fantastic opportunity, he believed: "Kurland and Livland . . . have been given that which the wide Russian land lacks—that is, the sea, the sea, the great road upon which the peoples of the world attain wealth, honor, and power. The wide Russian land on one side, on the other side the sea—oh, what a fine place!"[50] Yet this opportunity was inexplicably being squandered:

> Neither our Baltic Germans, nor the Latvians, nor the Estonians have taken up the true and open path to spiritual and material development. Behind us to the east in Russia live a hundred million people, starting from Tashkent and Irkutsk in remote Asia, who all must send their goods through our hands, across the Baltic Sea, past our doors to the western countries . . . and we, Latvians, Germans, Estonians, to all of Europe's amazement, continue to sleep.[51]

Valdemārs urged enterprising Latvians and Estonians to break free from their peasant heritage and turn to commercial shipping and shipbuilding, a potentially more lucrative sector than agriculture and one relatively free from feudal constraints. Latvia's future, he was convinced, lay in the hands of those who *left the land*, for the sea was the ideal means by which the Latvian nation could transcend peasant status. Seafaring promised enrichment, entrée into new occupations, and exposure to foreign places and ideas. The geography of transit held the key to overcoming Latvia's marginalization, if only Latvians could recognize that "Kurland and Livland are in the very center of Europe."[52]

To facilitate the transformation of Latvian and Estonian peasants into wealthy seamen, Valdemārs undertook to democratize maritime education in the Russian Empire. At the beginning of his career, there were four naval academies in the empire, including two on Latvian territory, in Riga and Libau (Liepāja). Tuition was expensive, instruction was in Russian and German, and most of the students were from the cities. To develop the empire's commercial seafaring capacity, Valdemārs argued, the number of cadres needed to be greatly expanded, and peasants' sons from the coastal areas would provide an ideal source of recruits. To this end, many small naval schools should be established in coastal towns and villages, with low tuition, government scholarships, and instruction in local languages. This program was in fact adopted in the imperial government's 1867 law on naval schools, drafted by Valdemārs, who also helped establish the first school of the new type on the Livland coast at Ainaži in 1864. Forty-one schools were ultimately established in the empire, including ten in Latvia. According to a recent Latvian study: "We can say without exaggeration that Valdemārs was the second most important promoter of tsarist Russian seafaring after tsar Peter I."[53]

Just as other New Latvians were delving into Latvian history and folklore to develop the idea of Latvia as a historic nation with a unique culture, so Valdemārs drew upon history to construct an ancient seafaring identity for Latvians:

> The Latvian and Estonian peoples live on the very seacoast, and therefore even in the olden days of freedom they were hardy seafarers. As is known from the thirteenth-century historical accounts of Henry of Livonia, the ancient Couronians and Livs often attacked even the Danes and Swedes, although these were the most famed seamen in the world in those days. . . . Similar reports about Latvian and Estonian seafaring handed down to us by writers of other nationalities . . . adequately prove that both peoples in very ancient times were well-known seamen.
>
> This same high and proud status that the Latvians and Estonians had achieved in ancient times we should attempt to attain again in our day. Just as the Danes, Norwegians, Swedes in their merchant ships now travel all the world's seas and earn large profits, so should we, the Latvians and Estonians.[54]

Valdemārs was confident that the multicultural future he envisioned would not endanger the survival of the Latvian nation, as long as even cosmopolitan, multilingual Latvians continued to read Latvian literature and as long as most Latvians continued to live in the countryside, "in Latvian surroundings, where foreign languages are not needed, and where even if once learned, they would soon enough be forgotten."[55]

Stirrings of Peasantism

Through his promotion of homeland studies, Atis Kronvalds laid the discursive foundations for what soon became the dominant Latvian construction of nature as a repository of national history and ethnographic uniqueness. And Valdemārs articulated the rival trope of the multicultural crossroads, a trope that soon went underground after Latvian statehood was established in the 1920s, only to resurface after Latvia emerged from Soviet imperial confines. Neither the cosmopolitan Valdemārs nor the more inward-looking Kronvalds was much concerned with celebrating the virtues of peasant life; the former's seaward developmental vision was very much a "postproductivist" one, while the latter's ethnoscape was infused more with the distant heroism of tribal battles. Other national revivalists at the time did embrace rurality, however. The definitive expression of this conservative agrarianism was undoubtedly the 1898 novel *One's Own Little Corner, One's Own Little Piece of Land*. This short work by a provincial schoolteacher, Jānis Purapuķe (1864–

1902), came to be known as the "gospel of the Latvian landowning class and Holy Testament to the sanctity of private landed property."[56] Even in its own time, the book's title and theme "attained the status of a proverb throughout the whole nation."[57]

In this didactic melodrama, the hardworking farmhand Pēteris Zelmenis struggles through every imaginable stroke of misfortune (illness, fire, bad weather, invidious neighbors, malicious overseers, debt, and hunger) to build a prosperous farm on land rented from the manor lord. In his old age, having finally saved enough to buy the farm outright in the name of his only son, Juris, he exacts a deathbed promise that Juris will preserve the farm for his own sons. Juris soon incurs a crushing debt, however, thanks to the pernicious influence of his lazy, spendthrift, citified wife. Ultimately the farm is kept in the family, but Juris pays with his own life and that of his wife for failing to respect the sanctity of land. Only land, Purapuķe insists, can provide security, prosperity, and self-determination for oneself and future generations:

> Money is snow: under a hot sun, it melts and evaporates without a trace; land doesn't melt, wither, disappear; land is the repository of blessedness. Your labor, your sweat, and your tears are stored up in it and saved so as to later provide you your earned reward. No, land is sacred! Sometimes it tests a person's patience and endurance, but ultimately it never forgets to reward industriousness. . . . Land is sacred! and no truly upright person will go to ruin with land. What are you without land? Like a child without father or mother, a rolling stone; you don't even own the air you breathe, the water with which you quench your thirst. Can someone like that even speak of a homeland, of a fatherland? The homeland is another's property, and the fatherland is an unknown grave site.[58]

Before 1920, however, Purapuķe's rural conservatism won few adherents. While the New Latvians were inevitably of peasant origins themselves, their leading figures, as Andrejs Plakans observes, "were uncompromisingly urban in their orientation. The past for Latvians was only the story of their enslavement; the present offered two choices—the 'darkness' of rural life and the 'light' which was symbolized by the city, its professions, and its cultural institutions."[59] The dominant discourse of the National Awakening did not celebrate, but rather deconstructed the conflation of Latvianness with the land, advancing the counterhegemonic claim that Latvians were not essentially tillers of the soil.

The Awakeners were not only urban but also overwhelmingly cosmopolitan in outlook. Even the firmly nationalist Riga Latvian Association devoted considerable resources to translating important foreign works into Latvian, while the New Current of intellectuals and artists that dominated political and cultural life in the last years of the century rejected nationalism as an ideology in favor of Marxism. This is not to say that Latvian socialists were antinational: indeed, the New Current "took the existence of a Latvian nation for granted and used Latvian to express the 'new' ideas."[60] The Marxist struggle was not so much against capitalism in general as against specifically German oppression.[61] In their construction of Latvian identity, though, socialist intellectuals drew upon a discourse that was both international in origins and internationalist in values, and while they rejected German hegemony, they also envisioned a bright and prosperous future for Latvia within a multinational Russian Empire.

In the countryside, landless agricultural laborers and tenants greatly outnumbered landed peasants, and class tensions continued to mount not only between Latvians and Germans but also between the rural poor and the "gray barons," or wealthier Latvian landowners. Mired in poverty and wretched living conditions, most rural Latvians were drawn to socialism and radicalism, not to agrarian conservatism. Indeed, during the upheavals of 1905, writes David Kirby, "the epicenter of the revolution in the Baltic provinces was the Latvian countryside," and by 1917, "socialism had struck deep roots" there.[62]

The First Republic: Constructing the Nation of Farmers

The years of the Russian Revolution and World War I saw a struggle for power and influence between Latvian socialists and bourgeois nationalists, with popular support initially leaning heavily toward the former. In elections to Russia's constituent assembly in November 1917, Bolshevik candidates in the Latvian part of Livland won 72 percent of the popular vote.[63] This support was eroded, however, by the terror tactics—arrests and executions—employed by the Bolsheviks against both Germans and Latvians, and by the failure of the short-lived provisional Latvian Soviet government to satisfy the peasantry's land hunger. Soviet authorities refused to parcel out expropriated manors to the landless, instead proclaiming them state farms or renting them out to tenants, and they even nationalized peasant farms. In November 1918 the Social Democrats joined the nationalists in proclaiming an independent Republic of Latvia, with a provisional government headed by Kārlis Ulmanis, leader of the recently founded Farmers' Union Party. A Latvian army raised by the Ulmanis government, fighting under the slogan "land and freedom," made a tactical alliance with Germany and expelled the Red Army (and, with the backing of the British navy, some Russian White Army divisions led by the military adventurer Pavel Bermondt-Avalov) from all regions of Latvian territory by January 1920. Soviet Russia recognized the victorious Ulmanis government with a peace treaty signed in August.[64]

The new republic's political stage was dominated by the peasantism that swept through much of Eastern Europe during this period. Latvia and the other new nation-states carved out of the wreckage of Europe's empires confronted overwhelmingly rural electorates radicalized by poverty and land hunger. In response, many leaders embraced a peasantist ideology that transformed the nineteenth-century Romantic cultural celebration of rural authenticity into the basis of the political and economic order. These regimes, writes Ghita Ionescu, took "the individual peasant explicitly as [their] social prototype," employed economic nationalism to emancipate the nation from foreign domination, and claimed a political leadership role for the peasantry, "not only on account of its electoral preponderance but also because of its innate spiritual and national values."[65]

In Latvia the chief voice of peasantism was Ulmanis's Farmers' Union Party, which served in eleven out of thirteen governing coalitions, providing ten prime ministers and three out of four presidents. However, this prominence reflected the party's success in coalition building more than its electoral dominance. In fact, the Social Democrats were consistently the largest parliamentary faction, and parties representing ethnic minorities steadily increased their representation throughout the period.[66] David Kirby refers to an "inherent suspicion of big industry, foreign capital and an urban proletariat" pervading Baltic political culture in the interwar years,[67] but this suspicion was perhaps not so much inherent as rooted in electoral realities. Given the overwhelming support for socialism in general and Bolshevism in particular in the Latvian territories in 1917, neither the political success of bourgeois nationalism, nor even the creation of an independent Latvia nation-state itself, was in any sense inevitable.[68] From the outset, "Latvian political and cultural nationalists began to worry about a nation-state betrayed and planned to use the power of the state to preserve the nation."[69]

Because of wartime demographic shifts, the rural share of Latvia's population increased from 60 percent in 1914 to 76 percent in 1920. Nearly half of this rural population remained landless, and the government fully recognized the salience of land hunger in stoking the fires of revolution and Bolshevism. Large landowners were virtually all Baltic Germans who not only were associated with the repressive legacy of serfdom and aristocratic privilege, but also more recently had resisted the creation of an independent Latvian state and had maneuvered for colonization by Germany. Agrarian reform thus dominated the agenda of the new regime. Thanks to the acute tensions between the landed and landless, the reform was one of the most radical in Europe. All privately owned rural properties of more than 110 hectares were nationalized without compensation and redistributed, first to veterans of the national armed forces and landless rural residents, then to farmers who wanted to expand their holdings. Recipients had to pay for their land, but the government provided long-term mortgages with deferred payments.

Those who have studied the agrarian reform concur that its economic objectives—rebuilding Latvia's war-ravaged countryside and promoting agricultural development and modernization—took a backseat to the urgent political imperative of eliminating landlessness and building a conservative agrarian base for political legitimacy. As the minister of agriculture at the time later recalled: "It was only the promise to satisfy the land hunger of the landless that made it possible to form an army, raise its moral quality, and ensure victory over external and internal enemies and the very creation of independent Latvia."[70] If feudalism had made the Latvian "a stranger in his own fatherland" (as Kronvalds and Valdemārs put it), then the goal of the agrarian reform was, in the words of one author, "for all Latvian farmers to feel at home" by becoming landowners (and, by extension, for all Latvian citizens to feel at home in their new state).[71] By this measure, the reform was a resounding success: it nearly doubled the number of agricultural holdings, thereby creating a large class of new homesteaders and reducing the ranks of the landless to less than a third of the prewar high. All told, 280,000 people, or 15 percent of the Latvian population, became new landowners.[72]

The designers of the reform chose to favor smallholders, forgoing preservation of even the more efficient of the large manorial farms. According to Arnolds Aizsilnieks, an émigré historian, Latvian legislators never seriously discussed the impact of farm size on productivity and efficiency. Driven by political rather than economic motives, they "considered as a self-evident ideal the so-called labor farm," that is, a farm small enough to be cultivated by a single family with the labor of its own members and their livestock.[73] When the reform was completed, farms of twenty hectares or less accounted for nearly 60 percent of all farms (but just over a third of all farmland), and farm-owning families made up four-fifths of the agricultural labor force. In 1935, 85 percent of all farms were worked by their owners.[74]

In short, the agrarian reform realized a vision of Latvian farmers as not only landowners, but *laborers on the land*—a vision voiced explicitly in 1920 by a delegate to the new republic's Constitutional Convention:

> Latvia's natural conditions and land features and the character, inclinations, and work traditions of the Latvian people indicate that the foundation of Latvia's economy and the welfare of the Latvian people has been, is, and will long continue to be agriculture. Fields, meadows, and forests are Latvia's only gold mine, and it is the work ethic of our laboring nation that has given and will give us the opportunity to most fully exploit this mine.[75]

Promoters of smallholding sought not only to create as many new landowners as possible, but also more specifically to valorize the personal labor of landed farmers. To ensure the consolidation of a conservative, capitalist, ethnically homogeneous nation-state, a new class of landed farmers had to be created, but one that would serve as the prototype of Latvian national identity. By implementing the vision of Latvians as an agrarian "laboring nation," the nationalist government reclaimed the politically charged realm of labor from Bolshevism and gave it a safely agrarian content.

The nationalist significance of agrarian labor was made manifest in the Freedom Monument, a sovereign symbol of Latvia's independent nation-statehood, unveiled in the center of Riga in 1935. Among the figure clusters in the sculpture is a group called "Work." As Daina Stukuls Eglitis observes: "At its center is a powerful figure clutching a *gudrības zizlis* (wand of wisdom) with a cascade of oak leaves spilling down from it: he represents the Latvian farmer; . . . the farmer holds a position equivalent to the Latvian warrior who stands, sword in hand, on the other side of the inscription, For Fatherland and Freedom."[76]

Another prominent site for reinforcing the agrarian national vision was a new open-air ethnographic museum, conceived in the early 1920s and opened in 1932, celebrating Latvia's peasant heritage. The museum featured buildings transported from actual Latvian farmsteads. Aldis Purs points out that these buildings, though ostensibly representative, were consciously selected to project the "*proper* national identity and a doctored view of Latvian history. Each of the four homesteads were individual smallholders, not communal peasants or landless laborers for a Baltic lord (probably more representative for many Latvian peasants in the nineteenth century)." The museum thus constructed the individual smallholder's isolated farmstead—the *viensēta*—as the essential type of Latvian agricultural life. In contrast to the traditional Russian commune and the fortified villages of Estonia, most peasants in the western and northern portions of present-day Latvia had lived and labored on dispersed farmsteads since the eleventh and twelfth centuries.[77] But in the eastern regions of Latgale and Augšzeme, which until independence belonged not to the Baltikum but to the Belorussian province of Vitebsk, the peasant commune had been the dominant form of settlement and land tenure. The Latvian land reform of the 1920s sought to liquidate the communes and move farmers to *viensētas*, and the new ethnographic museum omitted communes entirely as foreign and anomalous. Moreover, notes Purs:

> The museum had only Latvian homesteads, as if there were only Latvian peasants, and only Latvians in the state. Roughly one quarter of the population, however, was not ethnically Latvian, and in Latgale the share approached forty percent. Although

most minorities lived in the cities (particularly Riga), in Latgale most lived in the countryside. The inter-war museum had no Russian farm (communal or not), no Polish homestead, no gypsy compound, no Jewish stetl, no Baltic German manor house, no Belorussians, Lithuanians, Estonians or Livs. Instead the museum displayed a slice of ethnic agricultural life as if Latvian peasants existed in a world of their own, a world self-enclosed by the borders of the state.[78]

The ethnographic museum, in short, represented a conscious repudiation of Latvia's urban, commercial, and multicultural heritage in favor of the notion that "Latvian identity was peasant identity." This ethnically and religiously homogeneous agrarian vision, reified in the form of the *viensēta*, was presented "as a mirror on to the 'ancient identity' of Latvians (in the place of grand architectural monuments that Latvians did not have)."[79]

The agrarian Latvian identity was buttressed in interwar belles lettres as well. Literary "positivists" offering nationalistic portrayals of rural life and labor found common ground with nationalist ideologues and proponents of the newly invented "ancient" Latvian pagan religion, Dievturība, in affirming "the necessity of cultivating everything Latvian (in daily life, education, culture) and keeping away from all expressions of cosmopolitanism." For positivists, "The optimistic spirit of the times was most vividly manifested in the work of rural renewal and construction, in the reclaimed land, which had finally fulfilled the nation's centuries-old yearnings and raised its self-confidence."[80] Provincial theaters produced countless plays depicting rural life, and novelists reasserted the antiurban values of nineteenth-century agrarian conservatives like Jānis Purapuķe, whose *Little Piece of Land* was lauded by interwar critics as a classic expression of "Latvians' great striving for landed property."[81]

Purapuķe's theme was revisited by the hugely popular writer Jānis Jaunsudrabiņš (1877–1962) in his allegorical novel *The Homesteader and the Devil* (1933). The effete, cynical, and corrupt denizens of Riga provide the foil to Jaunsudrabiņš's vigorous and patriotic homesteader, a revered officer and wounded veteran of Latvia's war for independence who, for patriotic reasons, had traded his comfortable city life for the grueling ordeal of starting a farm. "I was driven here by concern for the future," he declares. "Concern for the future of my child, of the nation, of our state."[82] Like Purapuķe a generation earlier, Jaunsudrabiņš saw landownership and agrarian labor as the sole bastions of prosperity and freedom for both the individual and the nation:

> There were nonetheless people, a whole gray army of people, who believed in their state, because they believed in their work. Generations had yearned for their own piece of land, and now all the landless, the rural and even city workers were reaching

for it with both hands. The land, which the Latvian had trodden as a slave for long centuries, was now liberated and belonged once more to the nation. The joy was great, and in the first place was the joy of work. . . . It is true, many had nothing more than their hands, but then for a Latvian, hands mean wealth.[83]

As for Purapuķe, for Jaunsudrabiņš the farmstead or *viensēta* was the central embodiment of this ethic. Both authors rhapsodically depicted the process of constructing the new farm buildings, and for Jaunsudrabiņš's hero, moving onto his new land was "some kind of ritual, some uncontrollable internal drive."[84]

Jaunsudrabiņš's agrarian epic never gained much favor among contemporary readers or critics, perhaps on account of its overly "rosy enthusiasm."[85] Nonetheless, despite their questionable literary status, both Purapuķe's *Little Piece of Land* and Jaunsudrabiņš's *Homesteader and the Devil* were among the canonical works chosen for publication immediately after the war by the impoverished Latvian exile community in displaced persons camps in Germany, thereby facilitating the transmission of agrarian nationalist discourse beyond the geographical and temporal boundaries of the Latvian state.

Constructing the Agrarian Ethnoscape

When Atis Kronvalds began promoting homeland studies in the 1860s, Latvians had only recently been emancipated from serfdom and still chafed under German cultural and economic dominion. For Kronvalds, the salient images from the national past were those of tribal fortresses and ancient conquests. Latvians of the interwar period followed his lead in viewing nature as a fount of patriotic love, a reservoir of national character, and a "living witness" to the nation's past. But in keeping with the peasantist spirit of the times, they found significance in the geography not only of a distant and hazy golden age of heroism but also of the more recent and humble agrarian past.

The centrality of the agrarian ethnoscape in the cultural imaginings of the new republic is evident in many of its major literary works. Latvian intellectuals were increasingly urban dwellers themselves, yet most were scions of farm owners, farmhands, or rural artisans, and many canonical prose works of this period manifest a powerful current of nostalgia for rural life and the farmstead. A popular genre was the idyllic agrarian childhood memoir. *The White Book* (1927), by Jānis Jaunsudrabiņš, is a collection of sketches illustrating the author's childhood as a farmhand's son and remains today his most beloved work (and one vastly better received than his *Homesteader and the Devil*). The opening sketch, "Our House," is a loving and richly detailed description of his homestead in its entirety: the dwelling house, the garden, the barns. The sketch powerfully conveys closeness to nature, but a nature experienced through the *viensēta* in which land, people, and buildings are seamlessly intertwined: "All around our house there were fields, groves, forests, hills. . . . All of these waved to me and invited me. . . . Hail to you, you gray sand, which once received my footprints in your soft hands! Hail to you, you dear people, who so often steered my steps and my thoughts to the good! Hail to you, you old buildings, where I once dwelled and warmed myself!"[86] The nostalgic agrarian romanticism of Jaunsudrabiņš's description of the crowded dwelling house, where owners and farmhands together slept, cooked, stored food, hunted cockroaches, and watered the chickens in four small, dirt-floored rooms, is a striking contrast to the pathetic description of a typical Latvian *viensēta* penned more than a century earlier by Garlieb Merkel, the Herderian critic of serfdom:

> Even today the peasant dwellings in Livland are dispersed, often entirely isolated in thick forests. Usually these are barns or thatch-roofed huts without fireplaces or windows and with such low doors that one can enter only by stooping. The smoke-filled dwelling room is full of activity—the owner and his family, the farmhands with theirs go about their business by torchlight, and right beside them are the chickens, pigs, and dogs; . . . the children wear the same shirt in summer and winter and are always barefoot. But their attire is not the saddest part. Look into their faces! The transformed, bleak faces sneeringly tell you of hunger, cruelty, and the cowardly labor of slaves.[87]

While Jaunsudrabiņš's childhood home has windows and a chimney, the *viensēta* had otherwise remained unchanged in many respects since the days of serfdom. Yet whereas for Merkel it symbolized feudal oppression, now it represents the agrarian idyll.

The childhood sketches of Jānis Akurāters (1876–1937) share Jaunsudrabiņš's romantic merging of self, nature, and farm, but focus more directly on agricultural labor. *The Young Farmhand's Summer* (1907) traces the seasonal rhythms of farm work with a Tolstoyan interweaving of physical hardship and spiritual transcendence. Akurāters likens sowing time to "some religious cult" and writes of a fellow farmhand: "You are a shaper of the earth, second after God, your creator." Agricultural labor is seen as an integral element of nature itself, and also as the redemptive fulfillment of an intrinsic human need:

> In the mornings the entire wide clearing, as far as the eye can see, is full of hay mowers and girls and women. The mowers turn like white birds in a row and bend in rhythm—as if they were swimming on the high grass. The scythes sing from end to end. The girls move about in light cotton clothes and white kerchiefs. Somewhere there is singing and rejoicing, somewhere horses neigh, and someone hammers at barn roofs. Everyone is cheerful and happy, and it seems that everyone needs work.

And people observe each other from a distance, and the sun and sky embrace them with their summer hands.[88]

The memoir was written in 1907 while Akurāters, a social democratic activist, was in exile in Norway, and in the opening scene, exile from Latvia is merged with exile from the countryside:

> My window has been grown over by a great stone building, and every morning when I awake, I see its bony back and sooty window eyes. The sun is who knows how high above the rooftops, and only the stench of smoke and the barking noise of closing gates and doors enter my room.
>
> I glance at the walls: even those are of stone and exude cold even on summer days. And I ask myself: where am I? And it seems to me that I am forever somewhere strange; I have once left my homeland and can no longer return to it.
>
> My homeland? Yes, what then is my homeland? Dreams enclose me in their hands, and I see my homeland. The bird-cherry trees give off their fragrance, and I am once again walking through green valleys and listening to the corn-crake calling at the edge of the wheat field.
>
> And the expanse opens great and endless, and I am there, where winds flutter and white skies sleep at midday, and trees buzz with bees, and lindens bloom on overgrown riverbanks.
>
> And a young farmhand in gray clothing walks along the edge of the field, and I greet him. . . . it is I.

For Akurāters, the city is exile, homeland is the agrarian countryside, and the farm laborer is the unalienated self and citizen.

Akurāters's themes were echoed in the late 1920s by Anna Brigadere (1861–1933) in her classic three-volume memoir, *God, Nature, Work* (*Dievs, Daba, Darbs,* 1927). The first volume is revered for its vivid depictions of nature conveyed through Brigadere's recollections of tending herds at pasture—perhaps the defining childhood experience for Latvians even today, including city folk with relatives in the country. Its title has become a national icon, like the book itself. The third volume, depicting Brigadere's years in a provincial town, is called *In a Stone Cage,* for "the author wanted to emphasize that her true life is in the countryside, with which she has been intertwined since childhood."[89]

The work perhaps most frequently invoked by Latvians today as representing an essentially Latvian relationship to nature[90] is a drama, *Indrāni* (1904), by Rūdolfs Blaumanis (1858–1929), a farmer's son and another of Latvia's literary giants. The plot of *Indrāni* revolves around the dilemma of the old farmer, Indrāns, who reluctantly agrees to pass on the Indrāni farmstead to his malicious older son after the latter tricks him into believing that the good younger son will not return from Russia after completing his military service. The tragic climax comes when the new owner cuts down old Indrāns's beloved ash trees, planted decades ago around the farmhouse, and sells them to the local carpenter. The outraged father, before expiring in the outbuilding to which he has been exiled, denounces the son's narrow utilitarianism and explains the significance of the felled trees:

> Seed can be sowed in a field anywhere in the world. Land remains land, whether it's in [Latvia] or Estonia. But that clay hill there at the edge of the glade, that little stream in the sauna ditch, that rowan sapling here by the roadside, these belong only to In-drāni. Level that hill, divert that stream elsewhere, cut down that sapling, and you will have destroyed the soul of this house! Is this still Indrāni or some foreign place![91]

Indrāni is widely seen as the definitive literary embodiment of the Latvian conservationist sensibility, much as, for example, Uncle Vanya is invoked for Russia. But whereas Chekhov's Dr. Astrov is a scientific defender of Russia's vast expanses of primeval forests against large-scale logging, old Indrāns is the prototypical Latvian saimnieks: the good farmer who knows both how to tend his crops in the field and how to cultivate beauty in his farmstead. The threat against which the Indrāni ethic is mobilized is not the desecration of a primeval wilderness, but a disruption of the ordered landscape of the viensēta. When the Latvian self is exiled from this agrarian ethnoscape, he becomes a stranger in a strange land.

The Indrāni ethic was explicitly mobilized by the Latvian state between the world wars in a new annual event called Forest Days. Beginning in 1930, state foresters organized tree plantings in towns and villages across the country every spring. A central goal of Forest Days was to promote reforestation, since reconstruction after World War I had heavily tapped Latvia's forest resources. But an equally important aim was beautification of the Latvian cultural landscape through decorative plantings along roads, before public buildings, and in the barren yards of Latvia's new homesteads. In the lean postwar years, state foresters noted, Latvian farmers had become alienated from their traditional closeness to nature. The task of Forest Days was to reclaim this birthright:

> The Latvian nation, as a nation of farmers, has always loved nature and its most beautiful manifestation—the forest. . . . Love of individual trees and of the forest was expressed not only in folk songs and folktales, but in people's way of life. Houses were built at the forest's edge and around them were planted not only fruit trees, but also lindens, maples, oaks, and ashes, as well as other decorative trees and shrubs. These trees were planted not only for exploitation, but more for the sake of love. This love for the forest and trees is inborn in the Latvian nation, and we see it most vividly expressed in the figure of Blaumanis's old Indrāns. For him, every tree was precious. Each one was associated with some memories and events.[92]

In this postwar incarnation of Heimat protection, reinscribing the "traditional" Latvian identity by beautifying the *viensēta* went hand in hand with rehistoricizing the landscape by commemorating the memories and events most important to Latvian nation builders. Thus, trees were planted along the roadside where the revered foreign minister Zigfrīds Meierovics, architect of the international recognition of Latvian statehood, died in a tragic auto accident. Trees were planted at the sites of critical battles from the war for independence and along the Boulevard of Fallen Heroes. In addition to state foresters, the sponsors and organizers of these patriotic planting efforts included units of the Latvian Home Guard, war veterans' organizations, and the Boy Scouts and Girl Guides. Participation in Forest Days by individuals and state entities grew steadily from year to year but exploded exponentially after Kārlis Ulmanis's authoritarian coup in 1934 (to be discussed later).[93] In his opening speech delivered as "highest protector" of Forest Days, Ulmanis made explicit the role of beautification in cultivating love of the fatherland. He invited participants

> to bind themselves more closely to the land, to sink deeper roots into the land, so that the youth . . . might become aware of their future responsibilities and tasks as citizens from whom the future will demand passion, but also self-denial and sacrifice. Let them look back one day . . . and remember that somewhere a tree is growing whose roots tie and bind the planter himself to our land, to our precious land, bought with heavy sacrifices, which we must continue to love and for which we must fight and, if necessary, suffer and give ourselves.[94]

According to the 1939 chairman of Forest Days, participating with their labor in the beautification of the homeland allowed Latvians to "prove their loyalty to their country" in times of peace, just as twenty years earlier, "freedom fighters did so by defending it against enemies."[95]

Kārlis Ulmanis and the Golden Age of Agrarian Statism

The image of Latvians as an ancient nation of solitary, laboring homesteaders, enforced by the agrarian reform, rehearsed in literature, and reified in the ethnographic museum, represented a radical inversion of the nineteenth-century struggle to transcend peasanthood. Latvia's Valdemārian celebration of cosmopolitanism and commerce was further repudiated by privileging agriculture as the engine of national development and, conversely, by efforts to circumscribe the power of "big industry, foreign capital and the urban proletariat," in David Kirby's phrase. From the outset, Latvian governments invested significant resources in expanding agricultural capacity. Already in January 1920 the government charged the Ministry of Agriculture with ensuring the cultivation of all fallow agricultural land and began subsidizing farmers to achieve this goal.[96] During the interwar years governments used various policy tools to boost agricultural growth, including price supports, state purchases of grain, heavily subsidized timber for farm construction, loans in cash and seed, flood relief, and subsidies for field drainage.[97]

On the industrial and commercial front, Latvia, like many countries during this period, increasingly used state intervention to promote import substitution. "If we consider that the main policy of our country is to uplift a healthy peasantry and to avoid the growth of a factory proletariat," declared an editorial from the early 1920s, "it does not lie in the interest of Latvia to extend big industry, but rather only to facilitate such branches as are necessary for local requirements."[98] Foreign trade was to be controlled "so that Latvia would not serve merely as a bridge for foreign powers, which along with the bridge also trample the fate of the Latvians."[99] Editorials warned that "speculators must not be allowed to flourish at the expense of the majority of Latvia's residents."[100] The agrarian reform had established a preponderance of ethnic Latvians in agriculture, but ethnic minorities, especially Germans and Jews, continued to dominate industry, commerce, and many professions. Governments of the 1920s thus sought to foster the creation of Latvian-owned private enterprises. In the early 1930s, as Latvia reeled from the impact of global depression, these efforts meant "deep and far-reaching intervention into the entire economic life of the country."[101]

The debilitating effects of the depression, combined with a highly fractured parliamentary system, contributed to a growing sense of political instability. By 1933 industrial growth was already noticeably reviving, but in 1934 prime minister Kārlis Ulmanis exploited widespread perceptions of economic and political crisis as a pretext to seize power in a bloodless coup. Following the trend in many European countries during this troubled time, Ulmanis dissolved parliament, arrested political opponents, censored the media, and imposed one-man authoritarian rule over political life. The already extensive state control over the economy was tightened further through the establishment of a corporatist system of centrally controlled chambers modeled after Mussolini's fascist state.

Ulmanis—leader of the Farmers' Union Party, seven-time prime minister, and a revered founder of independent Latvia—was the ultimate avatar of the interwar agrarian nation-building project. Not only did he hail from a peasant family himself, but he was also a trained agronomist and a single-minded believer in agriculture as Latvia's developmental destiny. According to the memoir of a Latvian classmate, as a high school student in 1902 Ulmanis was interested only in agriculture: "Latvia? Latvia was a land of farmers, it needed to grow crops, fatten up good livestock, ensure good exports of butter, and so on."[102] As Latvia's leader (*Vadonis*; cf. *Führer, Duce*), Ulmanis promoted agricultural expansion through a range of subsidies and protective measures. In his own life and in his politics, Ulmanis embodied the very nexus of farming and nation building: "National unity, hard work, rural virtues, the belief that farming was the occupation closest to the Latvian soul and a necessity for Latvia's survival in the world economy—all these beliefs formed the basis of Ulmanis's pragmatic political philosophy." In short, "Ulmanis saw himself and was portrayed by his admirers as the *saimnieks* of Latvia," observes Andrejs Plakans, employing the semantically and symbolically rich word that can be translated variously as farmer, landowner, master, boss, or steward (cf. Russian, *khozyain*).[103] Despite the quasi-fascist character of the Ulmanis regime and the notorious linkages between ruralism and fascism in Nazi Germany, Ulmanis's ideology was not, in fact, fascist or racialist. According to Kirby, his "ideological world was based on a sentimental and idealized image of a sturdy, patriotic and patriarchal peasantry, a far cry from the dreams of racist power and conquest which inspired Hitler."[104]

While venerating the cultural landscape, the Heimat, in rituals such as Forest Days, the Ulmanis regime at the same time displayed a decidedly productivist orientation toward nature. In an era when the personal labor of the landed farmer was central to the project of nation building, land was viewed primarily as a productive resource. Government planners exhorted farmers, "Take as much out of the land as you possibly can, even though our land is fairly lean. The more we can learn to take out of the land . . . the better we will do."[105] Substantial state subsidies were provided for making nature more productive through land reclamation (drainage) and river regulation, and the state had the power to compel farmers to join reclamation associations.[106]

Indeed, reclamation was a potent symbol of the new homesteader's daunting struggle to eke a livelihood out of the war-ravaged countryside and, by extension, the Latvian nation's struggle to build a state. As noted earlier, Jānis Jaunsudrabiņš immortalized this struggle in his novel *The Homesteader and the Devil*, in which the heroic war veteran-turned-farmer expands his productive capacity by draining a bog inhabited by the devil (a common scenario in Latvian folklore), who has vowed to sabotage his epic homesteading efforts and return the land to its pristine state. "This ought to be good land," observed Jaunsudrabiņš's protagonist. "Perhaps the entire bog could be a first-class field, if the lake were lowered by some two meters and the peat dug out."[107] In 1936, the Ulmanis regime began a two-year project to drain 220 hectares of state-owned wetland between Riga and Jelgava. The drained area was christened Victory Field and divided into seventeen farms that were granted to war veterans. Thus in the nationalist symbolism of reclamation, life imitated art.[108]

The "Ulmanis Days" Reconsidered

Kārlis Ulmanis's performance as steward or *saimnieks* of the agrarian nation had a profound and enduring impact on Latvian popular consciousness. The dominant image of the interwar First Republic—especially the "Ulmanis days"—among Latvians today remains that of a nation of hardworking smallholders who exported butter and bacon to Western Europe. Indeed, the golden age imagined in post-Soviet Latvian national discourse is not Kronvalds's hoary era of pre-Christian heroism but the homely yet prosperous days of the 1930s, even though —as in many national myths—the image is of dubious historical accuracy.[109]

The postwar rebuilding of Latvian agriculture from the profound physical devastation of the war, which left much of the countryside a desolate moonscape of trenches, barbed wire, and ruined buildings, was undeniably a remarkable achievement and a powerful testimony to the Herculean labor of hundreds of thousands of new landowners. From these desperately meager beginnings, Latvian farmers were very quickly able not only to meet domestic needs but also to generate substantial exports of flax, butter, and bacon. Agriculture accounted for over a third of total exports throughout the period, and farming did play an important role in raising Latvian living standards nearly to West European levels.

But at the same time, to meet the great demand for land, the state supported the establishment of farms on poor land, and many new farmers lacked the experience and the capital needed for their daunting task. Small farms were chronically undercapitalized, less productive, and more burdened with debt than larger ones.[110] They were more labor-intensive, too, which exacerbated an ever-growing shortage of hired laborers on larger farms.[111] The heavily subsidized timber sales to homesteaders were tremendously costly and reduced state earnings in Latvia's largest export category.[112] Ultimately, as a Latvian analyst reported in 1940, agricultural productivity was significantly lower in Latvia than in Sweden and Norway, despite similar natural conditions.[113] Rather than an engine of national growth, in other words, Latvia's smallholders were arguably a net drain on national resources. The image of increasing prosperity itself, too, is something of a mirage: total and per capita gross national product actually declined between 1934 and 1937, making only a modest recovery in 1938.[114]

During the 1930s the unprofitability of agriculture, together with the lure of industry and urbanism, increasingly undermined the agrarian claim that "the foundation of the Latvian state is and will be agriculture and farmers."[115] Heedless of the farmstead's significance—in Ulmanis's words—as "the dwelling place of generations of farm families and the eternal source of the nation's living strength,"[116] growing numbers of farmers were provoking widespread alarm by abandoning their land. As an observer recalled:

> People were trying to get from the countryside to the cities, to do any kind of work. In some places old, well-to-do families died out completely. . . . In nearly every municipality you could find several farms where the fields were gradually growing over with brush and the buildings' roofs were threatening to collapse. The reason—the farmer and his wife are old, but the children have gone to the city in search of an easier life.[117]

Those farms that continued to function were beset by chronic and worsening labor shortages, and the government employed a combination of carrots and sticks to halt what it considered a very dangerous trend.[118] The problem nonetheless worsened as the decade came to a close, and increasing numbers of foreign workers were brought in from Byelorussia and other Soviet republics—a phenomenon Ulmanis campaigned against with the slogan "Be the tiller of your own land!"[119]

In short, the vision of Latvia as a nation of farmers was manifestly unraveling even under the most determinedly agrarian of regimes. Agriculture was glaringly unprofitable, employing 62 percent of the labor force in 1935 but yielding only 35 percent of gross national product.[120] Despite conditions of rural labor shortage and urban unemployment, rural Latvians were increasingly flocking to the cities. The sons and daughters of Latvia's homesteaders were abandoning the *viensēta*, the "eternal source of the nation's living strength," and despite the prevailing ethos of economic nationalism, their places had to be filled by foreigners.

Chapter 2

"The Occupation of Beauty"

In 1974 Imants Ziedonis, Latvia's poet laureate, posed a rhetorical question to his compatriots: "Will some collector or 'gatherer' appear today to fix the natural landscape of Latvia in its beauty as it still is today and, who knows, may not remain in the technical tomorrow?" Two years later, Ziedonis himself launched a quietly dissident movement to protect Latvia's pastoral landscapes from neglect and from the bulldozers of Soviet land reclamation brigades. To understand why the most important cultural figure in Brezhnev-era Latvia chose to spend most of his weekends for the next twenty years in the countryside "liberating" trees from undergrowth, we must examine the occupying Soviet authorities' assault on the "nation of farmers" and its agrarian ethnoscape. As the story of Ziedonis's Great Tree Liberation Group makes clear, Latvian responses to this assault were shaped by the discourse of Heimat—by the notion that Latvians become Latvian by shaping nature and that nature itself is a vessel of national history and character.

Unmaking the Nation of Farmers

The Ulmanis regime's attempts to stem the flow of Latvians from the countryside were cut short by World War II. Latvia's annexation by the Soviet Union brought a radical transformation of rural life and work through what Heinrihs Strods calls "three acts of violence against Latvian farmers": nationalization of land and attendant deportations, forced collectivization of agriculture, and liquidation of the individual homestead.[1] Nationalization began one month after Latvia's occupation by the Red Army and the capitulation of the Latvian government on June 17, 1940. In the first mass deportations from Latvia, on June 14, 1941, over 14,000 people were sent to forced labor camps in the Soviet far north and Central Asia. Three days later, Latvia's first collective farm (kolkhoz) was established.

Efforts to wipe out individual farmers began in earnest later in the 1940s. The Latvian pattern tragically recapitulated what had already happened elsewhere in the Soviet Union, following Stalin's fateful decision to "liquidate the kulaks as a class."[2] In Latvia as elsewhere, the category of kulak was defined broadly enough to include any moderately hardworking farmer.[3] Successful individual farmers were economically squeezed through increased compulsory deliveries of agricultural produce, denial of credit, and confiscatory taxation. Despite these pressures, Latvian farmers continued to resist collectivization in the familiar ways: abandoning their farms and fleeing to the cities, slaughtering their livestock in great numbers, and joining in armed partisan rebellion.[4] The decision to rapidly enforce mass collectivization, formally taken at the tenth congress of the Latvian Communist Party in January 1949, was given teeth through a second wave of deportations on March 25, 1949, which targeted over 43,000 "kulaks." By May 1, half of Latvian farms had been collectivized, and collectivization was completed by the end of 1950.[5]

While Soviet leaders believed in the economic superiority of huge-scale farming, the paramount aim of dekulakization and collectivization throughout the Union was political: eliminating the landed peasantry as a politically subversive class and transforming it into a rural proletariat of wage laborers at factory-style enterprises. In the newly annexed Baltic republics, too, collectivization was "intended to obliterate the social, economic and ideological foundations upon which the prewar independent republics had rested."[6] In Latvia, this political mission confronted a unique obstacle in the individual dispersed farmstead (viensēta) reified as the traditional embodiment of Latvian rural life that had served as a crucial symbol of national identity. As early as 1947, Soviet authorities began promoting construction of villages on collective farms. In an inversion of the 1920s bourgeois government's efforts to relocate communal peasants in Latgale to viensētas, beginning in the mid-1950s the Soviets sought to push farmers off of their homesteads into the new villages by banning new construction and restricting funds for repairs outside the villages.

The Soviet unmaking of the nation of farmers through collectivization and liquidation of farmsteads was aided by the unplanned depopulation of the countryside. As throughout the developed world, the rural proportion of the Latvian population shrank dramatically in the postwar period—from 65 percent in 1940 to 29 percent in 1985—though not for the same reasons as elsewhere. "If in our century the rural population of developed countries declined due to increases in agricultural labor productivity and overproduction," notes Strods, "then in Latvia people fled from the countryside."[7] War, emigration, and terror drained the countryside of hundreds of thousands during the 1940s and 1950s, and from the 1960s onward agricultural workers left for the cities in increasing numbers, among them three out of every four displaced viensētnieki who refused to move to kolkhoz villages. The combination of stagnant agricultural output and dwindling population

resulted in chronic shortages of both skilled and unskilled labor. Soviet authorities sought to alleviate the problem by drafting urbanites at harvest time in the
infamous "voluntary" Saturday work sessions (Latvian, talkas; Russian, subbotniki)
and by importing workers from neighboring Byelorussia and other Soviet republics. The Soviet Latvian government began recruiting in-migrants with cash pre-
miums, subsidies for housing construction and livestock, and other privileges.[8]
Through the combination of in-migration and deportations, the ethnic Latvian
share of the population dropped from around 76 percent in 1939 to 60 percent in
1953 and fell as low as 53 percent by 1985.[9] Rural laborers constituted only a small
portion of total labor in-migrants, but many state farms were established in areas
heavily depopulated by the war and consequently were manned largely by new-
comers. This influx further undermined the association of Latvianness with agriculture. "These farms lost the ancient work traditions characteristic of Latvian
peasant farms," laments Strods, and "Latvia's countryside lost its earlier face."[10]
In short, if in the 1930s the image of Latvia as a prosperous nation of smallholders
was beginning to unravel, in the Soviet period the image was exploded altogether.
Thanks to the Soviets' privileging of industry, agriculture accounted for an ever-
smaller proportion of national production and employment.[11] Latvians were no
longer landowners or viensētnieki, and increasingly they were not even agriculturalists, while the countryside was becoming less of a haven from ethnic hetero-
geneity.

Land Reclamation and Liquidation of the Individual Farmstead

In 1965 the Brezhnev administration inaugurated an ambitious campaign to in-
tensify land reclamation and agricultural mechanization throughout the western
and northern regions of the Soviet Union. In Latvia, the reclamation program provided increased funding and legitimacy for a more sustained assault on the family
farm. Thanks to Latvia's abundance of water (the amount of water per capita was
among the highest in the Union), drainage had been practiced for centuries. Simple open-ditch drainage in Latvia was "as ancient as agriculture," and the first
drainage regulations were enacted by Kurland's Grand Duke Jacob in 1665.[12] Despite Latvian governments' firm commitment to drainage during the interwar
years, however, the poverty of national budgets had allowed them to drain only
around 2 percent of the country's wetlands.[13] Under Soviet rule, however, a radically transformationist attitude toward nature was backed by sufficient funding to
implement a far more extensive drainage program.[14]

The extreme technocratic productivism of Soviet environmental practices has
been well documented. Grounded in the Marxists' promise of abolishing scarcity

through technological mastery of nature, as well as Stalin's subjugation of science to party decree, Soviet policy recognized virtually no limits on society's ability to manipulate nature in the pursuit of economic progress. This technological Prometheanism gave rise to a host of monumentally transformationist projects. Among these were the 1947 "Great Stalin Plan for the Transformation of Nature," the agronomic quackeries of Lysenko, Khrushchev's Virgin Lands scheme of the 1950s, and Brezhnev's grandiose land reclamation program, which included the infamous but never realized redirection of north-flowing rivers to support irrigation in the south.[15] In Latvia, wetland drainage was the most significant result of the Soviets' commitment to transforming nature. According to Juris Dreifelds, "During the 1960s and 1970s, drainage work absorbed about a third of all agricultural investments in Latvia. Indeed, Latvia, with less than one percent of the former territory of the USSR, claimed over 11 percent of all Soviet drained land."[16] The total area of drained wetlands increased more than sixfold from the interwar years to 1975.[17]

As it had been in the Ulmanis days, reclamation remained a potent symbol of productivism and progress in Soviet Latvia, and "swamplands [were] the one cultural landscape where the battle between nature and man [could] be seen most vividly."[18] In his Stalin Prize–winning novel *To the New Shore*, the Communist writer Vilis Lācis depicted another epic struggle against the swamp: this time, the devil's role was played by the reactionary bourgeois farmers who obstructed the Communist hero's campaign to achieve abundance by draining Snake Bog. Ultimately, "too great and mighty was Snake Bog for a little handful of people to engage it in battle. Like an age-old curse the bog lay in its enormous nest."[19]

Throughout the 1960s and 1970s, even contributors to the *Almanac of Nature and History*, a yearly publication of the Latvian Society for the Protection of Natural and Historical Monuments, toed the party line in support of reclamation, arguing that wetlands drainage would have no adverse effects either on ground and surface water levels or on flora and fauna.[20] The *Almanac* lauded one of the most dramatic cases of Soviet reclamation in Latvia—the radical transformation (through river regulation and construction of dams, polders, and fishponds) of the Lake Lubāns wetland ecosystem in eastern Latvia:

Farmers remember terrible tales of the angry Lubāns, which has always flooded surrounding fields, destroying crops and bringing ruin to country folk. It is said that people in the Lubāns area live poorly, in perpetual fear of floods; that large areas, namely wet meadows, are uninhabited and that only in good years do people come from all around to mow lush grass for their herds.

That is how it really used to be, but now, driving toward Lubāns, we no longer see farmers worrying about tomorrow and threatened by floods, or failed collective farms. Quite the opposite—now the Lubāns area has economically and socially flourishing large farms with especially active construction and intensive, all-around development. The Lake Lubāns area has been transformed beyond recognition.[21]

At the time of the Soviet Union's annexation of Latvia, agricultural lands comprised nearly 60 percent of the country's territory, and "rural landscapes were a mosaic of fields, meadows, forests, rivers, lakes and farmsteads." This mosaic agricultural landscape was a feudal legacy not only of "large manorial estates, with a characteristic complex of buildings, a park and tree-lined drives," but also of "peasant homesteads, with their gardens, tree clusters and scattered individual trees."[22] To create the vast, level fields required for large-scale drainage and mechanized cultivation, the elements of this mosaic landscape had to be destroyed: old field boundaries, groves, decorative tree plantings, and farmsteads. Contests were organized with prizes for the speediest relocations from homesteads to the multistory apartment buildings springing up in the new kolkhoz villages. In 1970 Latvian Communist Party officials announced a plan for the liquidation of all homesteads by 1985, declaring, "The relocation of citizens from private farms to rural population centers is one of the most important elements in the strengthening of socialist agriculture."[23]

Despite powerful sanctions, the response of farmstead dwellers fell far short of government expectations. Even so, nearly half of Latvia's farmsteads were ultimately destroyed, and although the *viensēta* was not eliminated entirely from the Latvian countryside, the shape of rural life was profoundly transformed.[24] The intimate relationship of the good farmer (*saimnieks*) to his farmstead (*viensēta*), in which exploitation and beautification were harmoniously balanced—the *Indrāni* relationship—was replaced by collectivized agriculture on a vast scale, and the *viensēta* itself as the primary embodiment of that relationship was leveled in the name of boosting productivity. The official reason for liquidating homesteads was to remove barriers to land reclamation, and yet documents suggest that reclamation was actually involved in only about half the cases.[25] What was at stake, clearly, was not only technological progress, but also the symbolic role of the *viensēta* as the embodiment of the "Latvian mentality." To entrench a new identity for Latvians as collective enterprise employees, Soviet authorities needed to eliminate the individual farm as the carefully cultivated "ethnographic link" to their identity as a nation of smallholders.

Ideologically orthodox Latvian commentators defended the liquidation of homesteads, arguing that "the homestead-type dwelling influences the farmers' way of thinking and working . . . [and] promotes detachment and independ-

Figure 1. The mosaic agricultural landscape with traditional farmstead (*viensēta*). Photo: Guntis Eniņš.

ence."²⁶ Wedding Marxist utilitarianism to the interwar Latvian discourse of the "laboring nation," contributors to the *Almanac of Nature and History* asked: "Which ultimately has greater aesthetic value—an old, uninhabited, and therefore untended farmstead, or a wide, uniform rye field that takes its place?"²⁷ Latvians' primordial closeness to nature was invoked, but with an emphasis on the transformation of nature through productive labor:

Latvian folklore and ethnographic data attest that a protective attitude toward nature is an essential expression of our nation's ethical and aesthetic culture. This attitude developed through the labor process, which over the centuries has been tied through unbreakable bonds to nature. . . . The more work is invested in nature, the closer and more lovable it becomes, and the more beautiful it seems. For example, to the farmer a planted field seems more beautiful than an unplanted one, and a mowed meadow more beautiful than an unmowed one.[28]

Great Tree Liberation

The late 1960s and 1970s were a time of nascent, or at least increasingly public, environmental consciousness in the Soviet Union, as elsewhere in the world. High rates of urbanization spurred growing interest in outdoor recreation, and scientists were beginning to sound alarms about the ecological impacts of many "projects of the century." Popular writers began to denounce the hubris of the technocrats' belief that, as Valentin Rasputin put it, "the earth is put together wrong and must be rearranged."[29] Rasputin and other Siberian writers mounted an influential literary campaign against the despoliation of Baikal, Siberia's magnificent freshwater lake, and its surrounding river basins through the heedless construction of dams and paper mills. These so-called village prose writers (*derevenshiki*) defended the moral, spiritual, and aesthetic values of nature and celebrated simple village life, replete with peasant wisdom and virtue.[30] Embittered by bureaucratic opposition to their efforts, these writers and other idealistic nature lovers began to question the Soviet system and to reject Soviet patriotism in favor of Russian nationalism. The primordial expanses of steppe and taiga were celebrated as shapers of the Russian national soul by neo-Herderian defenders of the cultural landscape, of Heimat. One was Vladimir Chivilikhin, a writer who came to view the battle against Siberian deforestation as a struggle for the survival of the Russian people. "The natural environment creates what, poetically, we call the soul of the people and in reality determines the salient characteristics of national culture," declared Chivilikhin. "The Russian character is impossible to imagine without [Russia's] expanses of forest."[31]

In Soviet Latvia, too, references to the aesthetic and spiritual value of nature began to appear in Latvian professional journals alongside paeans to bulldozed farmsteads and drained wetlands.[32] Growing demand for recreation led to the establishment in Latvia of the Soviet Union's second national park in 1973 (following Estonia's Laahemaa National Park, established the previous year) in the valleys of the Gauja River and its tributaries. From the late 1960s on, Latvians began openly to denounce threats to the cultural landscape posed by the drainage program. These objections were expressed cautiously at first, couched in acceptable Soviet terms of critique: irrationality, chaoticness, narrow departmentalism, voluntarism. In 1969 republican authorities responded to the public outcry with a decree mandating that reclamation work must also protect landscapes—including "individual valuable trees and tree groups with agro-technical, sanitary-hygienic, natural scientific, or aesthetic significance," as well as "all historically significant trees." Critics pointed out, however, that government decrees were often disregarded by farm bosses "who categorically demanded the removal of old oaks, tree lines, and exceptional trees."[33] In the mid-1970s, an idiosyncratic movement emerged in response to these threats led by one of the most important Latvian cultural figures of the period. Embracing the legacy of Kronvalds and Indrāni, this Great Tree Liberation Movement set out, in effect, to restore order to the garden and reinstate the exiled Latvian as steward of his individual farm.

Trees had figured prominently among designated monuments of nature in pre-Soviet Latvia. In the 1920s enthusiasts had begun compiling lists of trees that merited special attention, primarily trees that were "old and large, or notable for ritual, cultural, historical, or other reasons."[34] In 1962 the forest scientist Staņislāvs Saliņš published an article in a technical forestry journal that sought to revive the tradition of registering these so-called great trees (dižkoki).[35] At that time, only thirteen individual trees had protected status in Latvia. Revisiting the topic in 1970, Saliņš lamented that still only twenty-one great trees were legally protected, as compared to 330 and 236 in Estonia and Lithuania, respectively.[36] He warned that many great trees remained unregistered and, more important, that many were under threat from two Soviet-era agents of destruction: neglect and agricultural modernization. If these huge old trees, accustomed to growing in open spaces, were not freed from the encroachments of undergrowth, they would not only lose their aesthetic value, but also die sooner. More gravely, despite the 1969 decree on landscape protection, great trees were being destroyed in astonishing numbers by reclamation brigades. Stately oaks growing in the middle of a field, or beside farmhouses, or lining old manorial drives were destroyed by the thousands if they got in the way of a kolkhoz tractor driver or drainage engineer. If a tree was too large to be toppled even by several tractors chained together, it was blown up by a special explosion brigade.[37]

Saliņš's publications, particularly his *Latvia's Great and Rare Trees* (1974),[38] sparked a surge of popular concern. As a reader recalled: "Amateur enthusiasts strove to carry out the work that had been left undone by foresters. A whole era of searching out and rescuing great trees began, with fiery discussions in the press and work in groups."[39] By 1977, 635 great trees were under state protection, a number now far outstripping those in Estonia and Lithuania. Great trees had become the most numerous protected natural monuments in Latvia.[40] Two men spearheaded the popular battle to defend great trees against undergrowth and land reclamation. One was Guntis Eniņš, an essayist and self-taught naturalist who conducted a photojournalistic crusade throughout the 1970s and 1980s to raise awareness of great trees and expose their mistreatment and destruction. An especially inflammatory article published in 1978 provoked a flood of letters to the editor, leading to the establishment of a scientific commission that ultimately ruled against blowing up great trees.[41]

The other was Imants Ziedonis, the revered People's Poet and cultural icon who was described by contemporaries as "the burning focus of the Latvian perception of life"[42] and "the chief moral authority during [the Brezhnev years] in Latvia."[43] Ziedonis's passion for what might be termed homeland studies was a defining feature of his greatness. "Before Imants wrote a particularly influential and resonant essay about one of Latvia's provinces," notes a literary critic, "he crisscrossed it, like a fairy-tale king who walks through a mushroom forest, with a book under his arm and a magnifying glass to his eye."[44] In 1976, Ziedonis founded a Great Tree Liberation Group. With formal scientific legitimacy provided by Saliņš, this informal collection of writers, artists, doctors, teachers, and other mostly urban intellectuals made two hundred weekend excursions over a twenty-year period, traveling to rural locations to "liberate" great trees from undergrowth, restore old manor parks, and establish new decorative plantings—in short, to tend the neglected Latvian ethnoscape. In 1991 Ziedonis published a book chronicling the adventures of the great tree liberators, which shared with the writings of Eniņš and other great tree enthusiasts the common threads of Latvian agrarian Heimat protection.[45]

In these texts, nature was constructed as cultivated landscape, and the essential Latvian relationship to that landscape was identified as tending or cultivating nature. Cultivation connected Latvians not only to nature but also to the primordial roots of their identity as Latvians. Tending the landscape was the defining Latvian act, passed on from earlier generations and originating in the animistic religious practices of the proto-Latvian tribes. "Our forefathers primarily worshipped large trees (usually oaks)," observed Eniņš. "The tree for the ancient Latvian was the shrine, the altar with which he countered the church forced upon him by the Holy Order and the manor lords." Ritual practices associated with trees were preserved "for centuries, in spite of the manor lords and German preachers, who ordered these trees to be cut down or otherwise destroyed."[46] Indeed, according to Janīna Kursīte, pagan tree-worshipping practices such as offerings and sacrifices

Figure 2. Great tree liberator Imants Ziedonis. Photo: Gunārs Janaitis.

were observed in Latvia as late as the eighteenth century, and even in the nineteenth century German clerics battled against "superstitious" Latvians' refusal to fell sacred trees. Kursīte notes that the intergenerational conflict over tree felling in Blaumanis's Indrāni resonates with the Christian-pagan battle over sacred trees as links to ancient deities and totems. To this day, the memory of animistic practices is preserved in the many trees bearing the names Sacred Linden, Sacrificial Oak, and the like.[47]

The great tree also testified to the Herculean labor of slash-and-burn farmers of yore who carved their fields out of the primeval forest. For Eniņš, great trees signified the very opposite of the forest: "Looking at a map, we see a connection: where there's more forest, there are fewer great trees. Might not the number of great trees be a peculiar indicator of the level of culture? It is humans who decide the fate of trees and shape the face of their land."[48] How did Latvia come to be a "great tree wonderland," Eniņš wondered, where so many trees could grow to such great age and size at such a northern latitude? His völkisch explanation was rooted in the unique Latvian relationship to nature: in the "deep piety and particular reverence of the Latvian toward trees since the days of the ancestors," in "the poetic soul and sense of landscape beauty of the Latvian peasant."[49] Latvian farmers of antiquity, he insisted, "protected not only sacrificial and sacred trees, but also every tree as an element of beauty."[50] Ziedonis concurred: "Latvians throughout the ages have lived in the society of trees and learned the languages of trees. . . . [They have] a special sense of tree hearing, inherited from the forefathers."[51]

Like the farmstead or viensēta, great trees were seen as reflecting the stalwart individualism that was considered a defining feature of the Latvian personality. As Eniņš put it in an interview:

> Everyone consciously or unconsciously, and usually more unconsciously, shapes his environment somehow in accord with his internal, deep archetypes, his internal essence. . . . Every homestead is like a reflection of the personality of whoever lives there. There are those who can get by without trees, but they are more likely Russians or maybe Lithuanians, whereas Latvians greatly loved to plant around their houses. . . . These trees personify the essence of the Latvian solitary individualist.[52]

Solitary great trees growing in open spaces mirrored Latvian farmers whose viensētas marked their difference from village-dwelling neighbors. Great trees were thus rich with not only cultural-historical, scientific, and aesthetic meaning, but also "national-ethnic" significance. For Eniņš, "These centuries-old trees are our national pride, because they form the unique face of Latvia."[53] In short, great trees exemplified the two-way influence of völkisch environmental determinism:

not only was Latvian identity constituted through the cultivation of nature, but nature itself came to embody that identity. "Clouds like this exist nowhere else in the world," observed Ziedonis wryly. "Sometimes it seems that clouds are a form of applied art and develop in response to the nation's taste and temperament. Sort of ethno-psychologically determined clouds!"[54]

In the Kronvaldian spirit, Eniņš and Ziedonis identified the great tree as a direct, living vessel of history: "This oak . . . was born around the time that Bishop Albert entered the land of the Livonians. The great tree has seen our ancestors in the salty sweat of feudal corvée; it was a witness to the terrible plague, when the land died out and a man could walk for several days without meeting another person."[55] Great trees were "like eternity itself," "monuments of nature and antiquity," "ancient, living witnesses to history."[56] When Soviet authorities destroyed cultivated agrarian landscapes, they simultaneously eroded historical memory. In Ziedonis's words:

> Here the Crusaders invited our leaders to negotiations, only to lock them in the castle and burn them. The undergrowth we cleared is burning in a huge bonfire. A terrible feeling. We know nothing about Latvia's history. We are not permitted to know it. The history of our people is being hidden from our children. Will we free ourselves from this undergrowth, from this gray mass that is swallowing up our culture-space?[57]

Great tree liberators sought to reclaim this lost knowledge by reviving traditional landscape cultivation practices and by historicizing nature itself. Thus they organized trips to the homes of important historical figures and named "liberated" trees after such figures, placing an engraved stone at the foot of each "living witness to history."

If the Latvian farmer, with his inherited sensitivity to landscape, exemplified the proper relationship between people and nature, then his opposite was the "migrant": the Soviet-era arrival from another Soviet republic, lacking a sense of place and of beauty because he lacked national identity. "This great country," wrote Ziedonis of the Soviet Union, "evolves slowly because the word *fatherland* has been destroyed in its people. Having once in revolutionary hatred rejected heritability, the tending of the fathers' work, it now suffers in its own stupidity."[58] Migrants were voracious consumers and destroyers with no sensitivity to nature, the antithesis of the *saimnieks*: "They have gnawed it all away! . . . They don't see, don't look, don't notice, don't hear the cries of the trees they have tortured."[59] It should be noted that Ziedonis attributed the migrants' insensitivity to landscape to their rootlessness, not their ethnicity. Migrants were destructive not because

Figure 3. Great tree liberators burning brush. Photo: Guntis Eniņš.

they had the wrong national identity, but because they seemingly had *no* national identity—because they lacked a sense of place, not because they had bad "blood." The chief threat to the Latvian ethnoscape was not the intrusion of a foreign culture so much as the erosion of national culture per se, through Soviet national assimilationism and the promotion of migrant labor to speed industrialization. The Russian village prose writer Valentin Rasputin criticized labor migrants in similarly anti-Soviet, rather than ethnic, terms. He writes in his novella "Fire":

> This was a bleak and untidy town, neither a city nor a village. It was more like a camp site, as if the residents, migrating from place to place, had stopped to wait out a storm and to rest up and had simply gotten stuck. . . . They seemed to be here just for the summer, and then just for the winter, never putting down deep roots, sprucing things up, or making improvements for the sake of their children and grandchildren. . . . Back in the old villages where they'd come from, people couldn't have imagined life without some greenery under their windows; here those same people didn't even have flower gardens on display.[60]

Thanks to decades of Soviet rule, Rasputin lamented, "even Siberians, especially city dwellers, have now grown lax in their attitude toward their native land, and they attack the autumnal taiga with as much greed as if tomorrow they will have to retreat or advance, as in a military campaign, rather than remain to the end of their days."[61] Both Rasputin and Ziedonis were following in the footsteps not of nineteenth-century German biological determinists or their explicitly racialist Nazi descendants, but of environmental determinists like Herder, who conceived of nature as determining national traits through the transmission of cultural characteristics associated with a particular place, rather than acquired "genetic" characteristics.[62]

For the great tree liberators, migrants represented one face of the larger catastrophe of Soviet rule, which had exiled Latvians from their own ethnoscape, leaving it untended: "Our land has no *saimnieks*. There are workers. But there is no *saimnieks*."[63] The Soviet campaign to eliminate Latvian farmsteads was a campaign against Latvianness itself, for it had eroded not only historical memory but also the characteristically Latvian sense of place: "Because homesteads are gradually being liquidated in our republic," wrote Mārtins Dāboliņš in 1980, "every year the number of local residents who know their immediate surroundings well, and who can report on natural and historical monuments in the area, is shrinking."[64] The technocratic, utilitarian Soviet relationship to nature had crowded out the primordial Latvian sensibility:

> The aesthetic of today's foresters is very practical: whatever is useful is beautiful. That's how we students were once taught about folklore: in the Latvian folk songs our ancestors only recognized the useful as beautiful. Those were lies. Several generations of Latvians were damaged that way. It got to the point of having arguments with drainage brigades about the beauty and utility of the oak. . . . "The flowering potato field is more beautiful than any oak!" And counterarguments were not heard.[65]

In the Aesopian language of Soviet dissent, the great trees symbolized Latvian victims of Soviet repression. Ziedonis warned that Latvians who had lost their sensitivity to trees had been "trained in the murder of small lives. So that when the time comes for big lives, your hand wouldn't tremble."[66] For Eniņš, the careless misuse of trees was "the occupation of beauty"; migrants' towns were "ghettos, concentration camps of crippled trees."[67] The great tree, on the other hand, was a "hoary soldier" and "guardian of the homeland, its branches spread wide as if blessing the earth."[68] And great tree liberation was a form of "soft dissent," in Ziedonis's words, not only against the technocratic utilitarianism of Soviet agricultural policy, but implicitly against Soviet occupation in general. Every kolkhoz, school, and village center should have its own great tree or group of trees, Eniņš suggested, as a place to "study beauty and love of the fatherland."[69] Ziedonis wondered:

> Will some collector or "gatherer" appear today to fix the natural landscape of Latvia in its beauty as it still is today and, who knows, may not remain in the technical to-morrow? The folk songs have been collected, so as not to disappear, but is not the beauty of our landscapes, homesteads, and roads also an expression of the nation's spirit, which, unfixed, may be lost to oblivion![70]

The weekend trips of the Great Tree Liberation Group were "permeated with some kind of ideal of the good," Ziedonis recalled. "It was undoubtedly patriotism. An unexpressed Latvianness. Some kind of instinct to preserve the values of one's people."[71]

The Ziedonis group was emphatically critical of official nature protection activities and rebuffed overtures of cooperation from the Communist Youth League, the Nature Protection Society, and other official entities. In Ziedonis's words, the

Figure 4. The occupation of beauty. Photo: Guntis Eniņš.

group "had to protect the naivety of positive thinking" against the cynicism of the "era of stagnation" by demonstrating that it was possible to care for something: "It may seem as if we are tending the environment, but it seems to us that the most important thing today is to tend oneself and the very idea of tending." "What's the point of tending an oak in the middle of the forest, where no one will see it?" asked Sarmīte Ēlerte, a former great tree liberator and now editor-in-chief of *Diena*, Latvia's leading daily newspaper. The meaning of the work, she argued, lay in the very fact that it was "pointless from the perspective of the system" and that by daring to define their own meaning, the participants were revealing the pointlessness of the system itself. Ziedonis compiled the book in the late 1980s from the individual and collective journals kept by participants throughout the years. Participants had exercised self-censorship in journal entries and public statements, Ziedonis told me, and words such as "patriotism" and "dissent" were not employed at the time. Nonetheless, the group's crypto-nationalist dissent must have been apparent to Soviet authorities, who in fact banned a similar group in Lithuania. According to Ziedonis, Latvian government and party authorities mounted no opposition to the group, thanks in part to his own privileged status as a decorated People's Poet.[72]

"Eco-nationalism" or Heimat Protection?

In the late 1980s, Mikhail Gorbachev's policy of glasnost (openness) unleashed a wave of environmental protest and organizing across the Soviet Union. In Latvia, the movement was ignited in late 1986 by an outspoken critique, published in a major cultural weekly, of a planned hydroelectric dam on the Daugava River near the city of Daugavpils.[73] This shocking exposé was followed by a series of editorials and roundtables, published in the same newspaper, in which many of the republic's leading scientists and intellectuals attacked the project. A public letter-writing campaign generated some 700 letters with 30,000 signatures, representing an unprecedented burst of grassroots dissent. Yielding to public pressure, the Soviet Latvian government convened an expert commission to investigate criticisms, and when the commission submitted an overwhelmingly negative report, the project was canceled in July 1987.

The anti-dam movement, like similar protest movements taking shape at this time in several other non-Russian regions, was laced with an anti-imperial rhetoric that linked environmental threats to misrule by Moscow. "Discontent [was] often voiced with the concentration of decision-making powers in Moscow. In addition, Moscow [was] portrayed as the destroyer of national cultures, historical monuments, and the like."[74] Newly formed environmental groups quickly spawned mass movements for autonomy and ultimately independence from the center. In Latvia, the anti-dam campaign spurred the creation of the Environmental Protection Club, many of whose members soon defected to such expressly nationalist organizations as the Latvian Popular Front and the National Independence Movement of Latvia. These groups quickly eclipsed their environmental progenitor in numbers and prominence. Dainis Īvans, coauthor of the Daugavpils exposé, became the revered first president of the Popular Front. After the breakup of the Soviet Union, the membership and visibility of the Environmental Protection Club rapidly dwindled, as did the mass environmental movements of the other breakaway regions.

Some scholars have concluded from this phenomenon of swift growth and demise that the mass environmental activism was less a manifestation of deeply held environmentalist beliefs than a mobilizational surrogate or safe proxy for deeper, but politically more dangerous, nationalist agendas. In this view, the primary concerns of these movements, aptly labeled "eco-nationalist" by Jane I. Dawson, were in fact not ecological at all but simply nationalist. Environmental threats, Dawson notes, were "viewed as a threat to the continued existence of a people or land."[75] In the Baltics, according to David Kirby, it was the grave demographic threat that brought "green issues . . . to the very heart of national existence."[76] Industrialization was resented not only for its own sake, but also as the engine of Russification. Most scholars have analyzed Latvia's anti-dam campaign in strictly eco-nationalist terms, as resistance to the prospect of increased inmigration of non-Latvian laborers. And, indeed, in Latvia and elsewhere, these movements brought about some stunning policy reversals, but they quickly shriveled to insignificance after 1991. Once national independence had been achieved and the economically strapped newly independent states had become responsible for their own national budgets, environmental movements lost their mobilizational utility and thereby also their mass support.

The notion of movement surrogacy is useful in explaining the rapid rise and fall of mass grassroots environmental activism across the Soviet Union and Eastern Europe. But the synchronic focus on the exceptional events of the late 1980s leaves out the stories of how discourses of nature and nation became so deeply intertwined over time. It therefore reaches the misleading conclusion that the nature-nation connection was *merely* instrumental and ephemeral, *merely* a short-lived and somewhat disingenuous response to the constraints of Soviet political space. In this view, environmentally unfriendly policy decisions of the newly independent states and the absence of continued mass environmental activism reflect low levels of environmental concern. It is hardly surprising, though, that economic concerns have outstripped ecological ones during the grueling years of post-Communist transformation. The dwindling of enthusiasm for street protests has not been confined to the realm of environmentalism alone; throughout the region and in every policy sphere, post-Communist politics has tended to "migrate from the street back to the government."[77] Nature protection may have receded from the dramatically visible arena of mass street politics, but it remains a site of intense activity among smaller circles of professionals and enthusiasts.

The eco-nationalism thesis explains the glasnost-era successes of the mass environmental movements in terms of their appeal "not only to popularly held environmental values but also to people's sense of national identity and community," and it explains their demise as proof of the greater depth and genuineness of the latter set of values.[78] Yet from the Heimat perspective, it makes little sense to oppose environmental and national values in this way. As we have seen, many Latvians since Atis Kronvalds have valued and protected nature precisely as a storehouse of national identity and cultural history, and national identity itself has been constructed in terms of nature protection: being a good *saimnieks* or steward of nature has been a crucial component of being authentically Latvian. The union of nature protection and national identity was not simply a marriage of convenience and certainly not, as Marshall Goldman claims, "an unlikely twist in an unlikely direction."[79] When Latvians of the late 1980s depicted threats to nature as threats to Latvianness, they were not simply using the language of nature as a code or proxy for the language of nation. Rather, they were reproducing a familiar discourse that constructs these two threats as essentially one and the same.

Opposition to the Daugavpils dam was very much a rejection of the center's imperial dictates. As one commentator put it, "Where the bear dominates, the hare drowns. Has anyone from all those planning institutes in Moscow ever once been in these places that they so casually want to bury?"[80] But it was also a rejection of particular land use practices. The overarching theme sounded in essays and letters by scientists, writers, and citizens from various walks of life was specifically the impending loss of a landscape of great natural and cultural-historical value.[81] The Daugava River became a flashpoint of resistance not as a generic symbol of Russifying industrialization but because of its rich significance as ethnoscape: because it is "an integral part of ancient mythology and cultural history," personalized in thousands of folk songs.[82]

The Gorbachev era was not the first time that Latvians mobilized to defend their "river of destiny." In 1958, during the earlier political thaw and brief flowering of "national Communism" under Khrushchev, the same newspaper published criticisms of an earlier dam project at Pļaviņas. While the author of this first anti-dam salvo carefully framed her argument in the ideologically acceptable language of economic rationality, she nonetheless noted pointedly that the ancient Daugava Valley was a protected natural monument encompassing "a whole range of historical monuments of value to the nation, including also the most ancient fortresses, which tell us about times long before our era." The author did not oppose construction of a dam per se, but rather a particular design that would entail flooding of the Daugava Valley with its three inextricably intertwined values: "beauty, natural riches, and historic places."[83] Symbolizing this loss most potently was the Staburags cliff, "a famous Latvian landmark . . . [that] with its perpetual spring had become a feature in Latvian myths, epic poems, songs, and books."[84] While public opposition succeeded in temporarily halting the project, in 1959 a purge of "national Communists" in the republic's leadership led to a renewal of the project and the flooding of Staburags. Today, a hugely enlarged photograph of the now-inundated cliff hangs in the Occupation Museum in Riga, a powerful symbol of "the occupation of beauty." In 1986, critics of the dam voiced the same consciousness—albeit now in more explicitly nationalist terms—of the Daugava as "the republic's national pride," of the river's "symbolic meaning in our nation's history and spiritual life."[85] They mourned the impending loss not only of habitat for rare species, but also of "characteristic farmsteads" unique to the threatened region and of the "bread from its rich arable lands."[86] Even a biologist emphasized how a natural landscape "as a symbol serves to support the nation's living spirit, its cultural circulation."[87]

In short, the threat posed to the ethnoscape by Soviet land use policies was experienced simultaneously as a threat to nature and to nation, and Latvians continued to mobilize against it even after the immediate danger of the Daugavpils dam was eliminated. In 1988 the Latvian Cultural Fund (an interwar organization renewed in 1986, thanks in large part to the efforts of great tree liberator Imants Ziedonis) launched its Daugava Program, aimed at "preserving and tending the natural, historical, and cultural heritage" of the Daugava. The program's mission included not only fighting against pollution and environmental degradation, but also fighting for

> renewal of the historically established, regionally characteristic rural and inhabited landscape, and the restoration of ancient, forgotten place names (of houses, streams, marshes, forests, meadows, roads, etc.) and their memorializing through sculpture and signs; the tending and renewal of old road networks, drives, and parks, inhabited landscapes, and monuments of history, culture, and folk architecture; [and] deeper and broader research in regional geology, geography, botany, zoology, landscapes, archeology, folklore, ethnography, and history.

To its founders, the Daugava Program represented a way of "regaining national self-respect, of knowing and renewing our fatherland."[88]

Post-Soviet Homeland Studies

In 1999, Latvian government authorities proposed siting a landfill near another landscape of great folkloric significance, the Blue Hill. As in the 1980s, cultural figures joined the Environmental Protection Club in protesting the desecration of this "symbol of Latvian national awakening and national liberation."[89] Even after the restoration of independence and the explosion of new demands on individuals' time and energy, the great tree liberators continued their weekend outings full force through 1995. Imants Ziedonis later became a top advisor at the Latvian Institute, founded in 1998 as a sort of national public relations agency charged with crafting and promoting Latvia's image in the international arena. Guntis Eniņš went on to administer the Latvian Natural Heritage Fund, established in 1998 under the auspices of the Latvian Academy of Sciences. Still answering the Kronvaldian call to "know the fatherland," the fund conducts regular expeditions to identify rare and unusual natural objects, including not only rare species' habitats and geological formations, but also "place names, ancient sanctuaries, petroglyphs, and the signs, myths, beliefs, and tales that ornament Latvia's hills, valleys, and forests."[90] Elected to the sixth parliament as a member of the radical nationalist For Latvia Party, Eniņš continued to campaign against the dual threats to nature and nation. He was awarded the prestigious Three-Star Order for his conservation efforts, but he declined it in protest against a "government that cares unforgivably little about the nation's spiritual values and about Latvianness."[91]

The conflation of nature with labor, history, and identity continued to be reinscribed in popular discourse. In the pages of the *Almanac of Nature and History*, the *viensēta* was rehabilitated as the defining element of the Latvian cultural landscape, reflecting the "order" resulting from "honorable and enduring labor": "The true Latvian cultural landscape develops around and within the farmstead, where the layout of the buildings harmonizes wonderfully with nature—with the hills and valleys or plains. Everything together corresponds to the character of our Latvian land."[92] Readers were reminded of the distinctiveness of the native farmstead as against the village settlements typical of neighboring Estonia, Lithuania, Russia, and Germany, and they were exhorted to tend and restore those homesteads that had weathered the ravages of time and Soviet policy. The comforting "psychological atmosphere of the ancient Latvian rural homestead" was analyzed as a factor of its spatial layout, and the vernacular architecture of the homestead was interpreted as a reflection of the "Latvian mentality."[93]

Under rubrics such as "The Latvian in His World of Harmony," contributors to the *Almanac* mined folklore for insights into the Latvian mentality. "We have a very ancient worldview," Latvian readers were reminded, "which allows us to survive in spite of everything. Let us preserve it, let us draw strength from our land and nature, and advice from the inherited wisdom of our fathers."[94] At the heart of this worldview is closeness to nature: "The word 'nature' in the contemporary sense does not appear in the Latvian folk songs. The only conclusion to be drawn is that apparently man did not feel separated from nature, and so there was no need for a special word."[95] "The Latvian is a protector of trees; he never dared to fell trees like the oak or linden without real need, and to destroy an apple tree was always a sin."[96] Even the grammar exercises in Latvia's new school textbooks, notes a Latvian sociologist, primarily employed "texts about rural life and nature."[97]

Latvians' special closeness to nature continued to be defined largely through the mediation of agrarian labor. A scholar reported in the *Almanac* that, according to the great majority of survey respondents, "Latvianness is manifested in a special relationship to the land (90 percent) and to work (80 percent)." Another identified "the work ethic, self-sufficiency, individualism, etc.," as topping the list of "eternal characteristics" of the Latvian people. For "at least a thousand years," contributors noted, the Latvian landscape had been shaped by the interaction of two forces: "man and nature, or more accurately, work and nature." A popular poet admonished her compatriots, "To call yourself a Latvian," you must "like a doctor observantly place your ear on the wide breast of the oak" and "your hands must become rough from labor."[98] Soviet oppression was characterized as the erosion of the essential Latvian role of *saimnieks*. Latvian history revealed that "the large farm mode of production is entirely incompatible with the Latvian character." Another writer protested, "It will be a long time before we remedy the destruction, before we turn from destroyers back into cultivators, before nature forgives us."[99]

A conference was organized in 1998 to commemorate the hundredth anniversary of the publication of Jānis Purapuķe's agrarian melodrama *One's Own Little Corner, One's Own Little Piece of Land*. A presenter noted:

> The resonance of Purapuķe's novel's title (combined with the work's substantive content) has been so powerful and enduring that it has been noted even by linguists. In the Latvian phraseological dictionary, "one's own little corner, one's own little piece of land" is entered under the heading "corner," and is illustrated with the use of this phrase in ten literary works published from the 1920s through the 1970s. A broader treatment of the theme is found in J. Mauliņš's 1989 novel *Underground Cur-*

rents, which depicts an honorable worker's battle for his living space and the difficulties of making a home in the urban environment, that is, Soviet Riga. One of Mauliņš's chapters is even titled "One's own little corner, one's own little piece of land."[100]

Politicians and government officials, too, emphasized closeness to nature in their constructions of Latvia's restored national identity. In the 1993 parliamentary elections, a campaign poster for the Conservatives and Farmers Party depicted trees in a field and chose as its slogan Anna Brigadere's now proverbial triad: "God, Nature, Work." Another depicted "a ramshackle country grain-threshing hut beneath the message 'The Latvian must become the keeper [*saimnieks*] of his own land!'" The slogan selected by the Latvian National Independence Movement was "Nation, Land, State," with a photo of "a majestic oak gently lit by the sun."[101] Both Anna Brigadere and Jānis Jaunsudrabiņš were commemorated on postage stamps in the 1990s. The Jaunsudrabiņš stamp depicted a scene from *The White Book*, which the Latvian Postal Service identified as "still one of the most popular pieces of Latvian literature."[102]

In most other post-Communist states, officials chose to depict famous individuals (either cultural and scientific figures, or rulers, in the cases of the larger historic nations) on the new banknotes. But the Latvians selected what the central bank calls "essential" symbols of "the Latvian environment and culture"—images associated with the agrarian ethnoscape: an oak tree, a panoramic view of the Daugava River, a "traditional Latvian homestead," a famous interwar image of a Latvian folk maiden and a portrait of Krišjānis Barons, "father of the *dainas*," the first great collector of Latvian folk songs. The only exception was the old sailing ship at sea—invoking Krišjānis Valdemārs's seaward developmental vision—depicted on the fifty-lat note.[103] Latvia's bankers also drew upon the ethnoscape in designing the new coins; the one- and two-lat pieces feature a salmon and a grazing cow, respectively, symbolizing "Latvia's abundant water resources" and "the wealth of the Latvian countryside."

The agrarian Heimat discourse continues to be transmitted from one generation to the next, as is evident in this "Letter to Society" published by a group of high school students in 1999:

> We have been given a wonderful fatherland—Latvia. There is neither gold, nor precious metals, nor coal in this land. Nor is it particularly fertile. One can live in it only through relentless labor. This demonstrates that Latvians, since prehistoric times, have not chased after instant riches, but we have made them ourselves, relying only on our own powers and labor.
>
> Moreover, the Latvian soul yearns for beauty, and most likely there is nothing

deeper in the soul than this yearning. . . . The sense of beauty and this demand have abided by him in the field, the meadow, the riverbank, the courtyard, the hearth.

These, it seems, are our nation's essential characteristics.[104]

Threats to nature continued to be perceived as threats to the nation, as Janīna Kursīte observed regarding Latvia's "river of destiny": "The Daugava has in our day become the dominant feature of the Latvian landscape, and the pollution or nonpollution of this river, the well- or poorly maintained character of its banks are linked both directly and indirectly in Latvians' consciousness with the nation's ability or inability to find its place in late twentieth-century Europe."[105]

The national significance of places like the Daugava basin was rooted in an environmentally determinist discourse that had remained remarkably true to its Herderian roots. Thus at the height of separatist fervor in 1990, Skaidrīte Albertiņa (an ecologist, independence activist, and Green Party deputy of the Latvian Supreme Soviet) offered this explanation for environmental degradation in Latvia:

> It is because populations have come in here from very rich and vast territories, where natural resources were very plentiful, where one could take as much as one wanted from the riches of nature. Living amid such great wealth, people developed a specific behavioral type, which I would call the hunter-gatherer type, because they were able simply to take and take. And then, lo, these people arrived here in Latvia, where we have only land, forest, and people—three riches of nature. . . . We are here on this little strip of land, and either we keep the land in good condition, so that it yields us the harvest, or else we are finished, for we simply have nowhere to go.[106]

Albertiņa's comments echoed the "frontier" thesis of the great nineteenth-century Russian historian S. M. Solov'ev, who saw Russia's vast open spaces as crucial determinants of national character and land use practices. Early Russia, according to Solov'ev,

> was an expansive, virgin country, which awaited population and awaited history. Because of this, the history of early Russia is the history of a country which is colonizing itself. Because of this there was a constant powerful movement of population across enormous spaces. . . . The settler does not remain long [in any one place]; as soon as he is constrained to work harder, he goes off to settle new areas, for there is open space all around, waiting to receive him; land property has no value, for the important factor is population.[107]

The Latvian discourse of nature and nation, like that of Solov'ev and of Russia's village prose movement a century later, was rooted in German geographers' environmentally determinist notion of the human cultural landscape, or Heimat. But it also differed from its Russian and German counterparts in important ways. For German conservationists, the formative experience of the late nineteenth century and early twentieth was that of breakneck industrialization and urbanization. For Russians, the perception of rapid natural resource extraction was balanced by that of limitless natural abundance, by the vastness of forest and steppe. For Valentin Rasputin, the sense of nature as intimately humanized ethnoscape was balanced by a sense of nature as sublime and awesomely nonhuman. Baikal "in all its stern, primordial beauty" was something "wrought by God," "alive, majestic, and not created by human hands, not comparable to anything and not repeated anywhere, . . . aware of its own primordial place and its own life force."[108]

Latvian constructions of nature, in contrast, are noteworthy for the absence of any notions of the sublime.[109] The Latvian discourse of Heimat developed in a context neither of rapid industrialization nor of natural abundance, but rather of the widespread hunger for farmland that was central to the experience of Latvian nation and state building. From the 1920s onward, the true Latvian was imagined not only as a hardworking smallholder but also as the tender of a natural landscape characterized more by scarcity than by abundance. Nature was neither a bountiful wilderness nor an industrializing landscape of modernization, but rather a garden tended intimately by the solitary saimnieks. As Guntis Eniņš puts it, the great tree was the very opposite of the forest.

Chapter 3

Globalizing the Ethnoscape

Shortly after the celebration of Latvian Independence Day in November 1997, the sociologist Tālis Tīsenkopfs published in Diena, Latvia's leading daily, an editorial entitled "Contemporary Latvia's Diverse Identities."[1] Tīsenkopfs cited an interview with Paul Goble, an American expert on the former Soviet Union who faulted Latvia for lacking a strong sense of identity or clear developmental orientation.[2] Whereas Estonians, in Goble's view, clearly identified themselves (and were identified) with Scandinavia, and Lithuanians with Poland and Central Europe, Latvians seemed to lack a vision of where they belonged or where they were heading. To the outside world, Latvia looked like a "blank spot," in Tīsenkopfs's words, poised between East and West. Tīsenkopfs concurred with Goble's assessment, conjuring the general ignorance of Latvia and Latvians with wry anecdotes from his travels in the West, where colleagues, when not confusing Latvia with Lithuania or the Balkans, "associate it with three things: oppressed Russians, Riga, and the Mafia."

Nonetheless, Tīsenkopfs saw a silver lining in this cloud of virtual nonexistence. "I like this blank spot on the map," he declared, "because it confers freedom and many choices for coloring it in." He noted that the 1997 United Nations Development Program's Human Development Report for Latvia had found that the country lacked a unified civic identity, but this very lack could be a great opportunity. It was time to shake off the nineteenth-century illusions of "ethnic nationalism," Tīsenkopfs argued, and embrace the diversity of identities itself as the "unifying principle for Latvia's constitutional and civic identity both internally and in foreign relations. . . . Why not unite on the basis of sound economic and legal values and develop Latvia as an economic and democratic space, viewed by the West with Hanseatic respect and by the East with Baltic wonder?" The beautiful capital city of Riga, Tīsenkopfs suggested, would be an ideal starting point for the construction of such an open, multicultural identity.

Coming just over a decade after the eruption of anti-Soviet Latvian nationalism under Gorbachev's policy of glasnost, Tīsenkopfs's interrogation of Latvian national identity was timely. On May 4, 1990, the newly elected Supreme Soviet of the Latvian Soviet Socialist Republic had declared the de jure restoration of the interwar Republic of Latvia. Full independence had been announced in August 1991, after the failed coup by Communist hardliners in Moscow, and Soviet troop withdrawal began in 1994. By the late 1990s, the former Hanseatic harbors were booming again, and Western investors had transformed the face of downtown Riga, where German, Swedish, and English could be heard almost as frequently as Latvian or Russian. The parliamentary political system functioned relatively smoothly, with multiparty contestation in free and fair elections every three years. With predominantly center-right, Western-oriented governments at the helm, Latvia joined the World Trade Organization in 1998 and became a formal candidate for membership in the European Union in 1999 and the North Atlantic Treaty Organization (NATO) in 2002. Ethnic Latvians dominated political and cultural life, and tough laws had been enacted to protect the Latvian language against displacement by English and Russian, the past and future idioms of business and power. In short, Latvia could claim a place among the most successfully consolidated new nation-states of the post-Communist region.

At the same time, however, corruption was endemic in Latvia, and, outside the Riga metropolitan area, economic growth was slow. International organizations had sharply criticized Latvia for an exclusionary citizenship policy that disenfranchised much of the large Russian-speaking minority population. As the heroic images of the Baltic "singing revolution" faded from memory, both Western media portrayals and scholarly analyses of Latvia tended to focus—like Tīsenkopfs's European colleagues—on the fraught relations between Latvians and Russophones. Meanwhile, internal Latvian debates on minority integration were haunted by the specters of demographic annihilation and a potential fifth column. While Latvia was well along the path of state building and democratization, in other words, the national character of this would-be nation-state, as well as its relationship to an expanding Europe on one border and an ambivalently postimperial Russia on the other, remained very much open to contestation. As Latvia struggled to chart its course amid the new pressures of globalization and the familiar pressures of its big neighbor to the east, Latvians were by no means certain about what it meant to be Latvian, and how to remain Latvian, in the new world order.

Not surprisingly, Tīsenkopfs's essay sparked an impassioned debate among Latvian intellectuals that simmered on the editorial pages throughout the ensuing months. Many commentators endorsed his vision of openness, stressing Latvia's historic role as an international crossroads, zone of cultural hybridity, and integral part of Europe. A historian declared:

From birth, the Latvian belongs historically, culturally, and geographically to Europe. Moreover, the blood of the German, Swedish, Polish, Jewish, Russian, and other nations flows in Latvians. The Latvian has gained from this—inherited many characteristics. For this reason the Latvian easily understands and assimilates the cultural and material values, the scientific and technical achievements, and ideas generated in Europe or America.[3]

Globalization was unavoidable, the historian warned, and Latvians must adapt to its flows. Another historian concurred: "In order to find our place successfully in the modern world, the people of Latvia must express themselves not solely in the categories of the national state. They must feel the link with processes that cross state boundaries."[4] Some embraced Latvia's once and future international identity without going so far as to reject ethno-national "illusions." Thus a doctoral student urged Latvians to feel pride as "a unique ethnos" and to celebrate the Latvian folkloric heritage, while at the same time he identified an innate tolerance of difference as a key marker of Latvianness. Latvians in the post-Soviet period needed to mend their tattered sense of "ethnocultural solidarity," he argued, but most of all, "We will have to act as switchmen, because our train—Latvia—has once again traveled onto international tracks."[5]

Others, however, rejected Tīsenkopfs's "postmodern" cosmopolitanism and economistic rationalism as a direct threat to ethnic and national identity. The esteemed author Miervaldis Birze pleaded for the defense of Latvia not as "an economic and democratic space," but as "the last and only refuge of the Latvian people" in the face of a demographic crisis of unprecedented proportions.[6] A theater critic celebrated nationalism as an instrument in the struggle against "cultural and economic imperialism" and an instrument of particular importance for a small nation under "postcolonial" conditions.[7] If for the internationalists the crucial symbol of Latvian national identity was the Hanseatic city of Rīga, then for their ethno-nationalist critics it was the countryside. "It is hard to imagine how the state can survive without the countryside," wrote the poet Anda Līce.[8] "Every city dweller's roots, after all, reach into the countryside of some Latvian district. For centuries the land has nourished the nation spiritually. . . . Does anyone, in speaking the word 'homeland,' imagine it only as the city?" Answering her own question, she quoted the novelist Arnolds Apse, who located the "soul of the homeland" not in the teeming polyglot streets of Rīga, but in the classic agrarian landscape: "A rain sonata on a gray plank roof. White birches on a lonely hillside. A threshing barn sunk into the ground, its roof covered with moss. An old willow by the side of the road. . . . The quiet fields, ripened for the harvest."[9] Echoing a theme pounded in the media and in political circles throughout the course of post-Soviet agrarian reform, Miervaldis Birze identified the countryside, with its ethnically more homogeneous population, as the wellspring of Latvian identity. Preserving Latvianness in a globalizing era, in Birze's view, would require not the pursuit of ever-greater openness and dynamism—English-language training, promotion of Internet access, and so on—but rather state support for family farms, rural schools, and Latvian culture.

As the philosopher Ella Buceniece observed, the Tīsenkopfs debate represented the latest installment in a struggle, ongoing since the National Awakening of the 1850s, between competing "models of historical self-understanding: the romantic or ethnocentric and the realist-rationalist or Eurocentric"—or, in my terms, between the discourses of agrarian nationalism and liberal internationalism.[10] Tīsenkopfs and his peers spoke the Valdemārian language of internationalism when they invoked Latvia's urban and maritime cosmopolitan heritage and celebrated openness to new ideas and transcending borders. Birze and the ethnoculturalists, on the other hand, reproduced decades of agrarian defense of Heimat in their calls for protecting the traditions of agricultural labor and the cultural landscape in which those traditions are inscribed. What was most striking about the Tīsenkopfs debate was precisely its faithful echo of themes and tropes of the 1860s, 1930s, and 1970s. Tīsenkopfs and his peers may have been free to choose how to "color in" the "blank spot" of Latvian identity, but the range of meaningful colors came from the two different palettes—internationalist and agrarian—that had dominated the history of Latvian nationalism.

What the Tīsenkopfs debate made clear, too, was the centrality of geographical imaginings—constructions of land and homeland—to the problem of defining the national self. The problem was how to define the distinctive features of not only Latvians but also the Latvian homeland. This chapter turns to battles over the fate of the agrarian ethnoscape—specifically, debates about land reform and agricultural policies—to see how discourses of internationalism and agrarianism structured Latvian visions of the post-Communist future.

By Land or by Sea: Debating Post-Soviet Rural Development

As we saw in chapter 2, the Soviet assault on the agrarian ethnoscape was successful on many levels. Farmers who survived war, deportation, and exile became kolkhoz employees, the countryside became less ethnically homogeneous, much of the rural mosaic landscape gave way to the flattened and drained tracts of large-scale mechanized collective farming, and half of Latvia's traditional isolated farmsteads, surrounded by decorative trees, were bulldozed. Yet paradoxically certain aspects of Soviet agricultural policy enabled a large number of Latvians to

retain an intimate relationship to the land. With enormous subsidies to the stagnant agricultural sector, the state propped up hundreds of unprofitable enterprises and protected thousands of agricultural jobs.[11] Wages for collective farm
workers in Latvia and its Baltic neighbors were well above the Soviet average.[12]
State support targeted farms in environmentally unfavorable regions, keeping
farming alive in places where, under competitive market conditions, farmers
would have had to pursue other livelihoods. While the rural share of the population dropped to only 30 percent, a significant decrease from the interwar period,
it was nonetheless very high compared to that of most industrial nations.

Moreover, in Latvia as throughout the Soviet Union, a curious dual structure
developed in agriculture. While collective farms were amalgamated to create ever-
larger enterprises, at the same time a parallel sector of tiny household plots allocated to kolkhoz and state farm (sovkhoz) workers flourished. These private plots
accounted for an enormous share of total production in the late Soviet period and
provided higher earnings than the average urban worker received.[13] Official tolerance of this individual sector allowed a much larger proportion of Latvians to continue farming—and to find it relatively lucrative—than would have been possible
in a market economy. The viensēta was bulldozed, in other words, but a vestige of
the owner-laborer's intimate relationship with the land survived in the form of the
household plot.

It is hardly surprising, then, that in the summer of 1991, over two-thirds of survey respondents predicted that agriculture would serve as the primary spring-
board for Latvia's economic recovery and development.[14] The 1990s brought bru-
tal economic dislocation and restructuring, however, and agriculture suffered in
particular. Because of the demise of the command economy and of Soviet and
Eastern bloc trade networks, production dramatically collapsed. By 1993 Latvia's
gross domestic product (GDP) had shrunk to half its 1990 level, and did not begin
to inch up again until 1996. The hyperinflation of 1992 was brought under control
by mid-decade, but a chain of bank failures in 1995 posed a grave setback for both
individual welfare and macroeconomic stabilization. Unemployment, which had
begun to decline after 1995, ballooned again as a result of the August 1998 economic crisis in Russia and the global recession of that year, which exacerbated
Latvia's growing trade deficit. Toward the end of the first post-Soviet decade,
economists concluded that while a small minority had profited greatly from
Latvia's transition to a market economy, the majority had endured a "dramatic increase in poverty [and] social inequality."[15]

With a series of laws enacted after 1990, the resurrected parliament launched
an agrarian reform. In the spectrum of post-Communist reforms, those of Latvia

and its Baltic neighbors represented the most radical commitment to restoring the former agrarian structure. First, all three countries dismantled the collective sector more aggressively than elsewhere in the region, and Latvia was the most aggressive of all. By 1998, reconstituted collective farms accounted for only 11 percent of Latvian farmland.[16] Second, all three countries elected to restore landed property rights directly to pre-Soviet landowners or their heirs, and Latvia and Estonia sought to restore to their owners full parcels of land "in the old borders," without regard to the use rights of Soviet-era tenants or demands for distributional fairness by former collective farm employees.[17] The majority of former landowners opted to reclaim their land, rather than receive monetary compensation, and many new property owners—newly transplanted city folk as well as former kolkhoz employees—chose to become farmers. In Latvia, over 64,000 private farms were established by 1995.[18]

Of course, the return to family farming in the Baltics can be explained in many ways. The choice of land over monetary compensation was influenced by the low market value of the latter, and the flight from city to farm was driven not only by a desire for restitution but also by "acute food shortages, unemployment, [and] economic stagnation in the cities."[19] Along with economic incentives, however, the power of agrarian nationalism clearly played an important role. An explicit aim of the Latvian reform, as stated in the 1990 land reform law, was "renewal of the traditional Latvian way of life."[20] "The national-cultural image of the free hardworking peasant farmer was indeed the key motivation of the agrarian reforms in the Baltic in 1990–92," claims Anatol Lieven.[21] According to Tīsenkopfs,

> One-half of the new farmers [in Latvia] were inspired by a desire to restore the farm that their parents or grandparents once ran; one-quarter were forced into individual farming as the only possibility to maintain a livelihood after the collapse of the kolkhozes; and only a minority of eight per cent wanted to start up a farm as a business. The agrarian reform, which started as a political act, thus took on the characteristics of a historical and cultural movement.[22]

In the early 1990s, the question of private landed property was "even more emotional than it was during the First Republic," an agriculture ministry official told me. "The landowning spirit is so strong in us that it was not destroyed in fifty years [of Soviet rule]."[23]

Yet most of Latvia's new farmers lacked the necessary capital, equipment, and know-how, and their landholdings were too small or fragmented, to function effectively under market conditions. Caught in the "scissors" of declining agricultural commodity prices and rising input costs, farmers suffered sharp income losses in the second half of the decade.[24] Huge disparities began to develop in the size and prosperity of farms, with a few rapidly growing large farms in the fertile central region on the one hand, and a large percentage of small farms, many reduced to subsistence production, on the other.[25] The Baltic Free Trade Agreement of 1997 eliminated all tariff and nontariff barriers among the three Baltic countries, but the failure to harmonize agricultural support policies and external trade policies sharply disadvantaged Latvian producers. The situation was exacerbated after 1998 when Latvia joined the World Trade Organization but failed to enact permissible protectionist measures such as an antidumping law. Unable to compete with imports, the contribution of agriculture to Latvia's trade balance plummeted from a net surplus of 52.8 million lats in 1993 to a 134.5 million–lat deficit in 1998, and agricultural production fell to 43 percent of its 1990 level.[26]

The decline in agriculture was particularly dramatic in comparison to other sectors. The contribution to GDP of the agrarian sector (agriculture, fishing, hunting, and forestry) dwindled from 21 percent in 1990 to 4.5 percent in 1998, while that of services doubled from 32 to 66 percent.[27] Farm wages remained well below the national average, and rural unemployment considerably outpaced urban levels.[28] Latvia's postindependence governments, dominated by center-right parties, were loath to prop up the agrarian sector.[29] A Law on Agriculture in 1996 mandated that at least 3 percent of the annual budget be allocated for agricultural subsidies, but this slight increase was inadequate to offset the costs of restructuring and market liberalization for most farmers. In short, as Tīsenkopfs observes: "Despite the fact that rural ideals, along with nationalism, greatly inspired the restoration of Latvia's independence, the countryside turned out to be the major loser in the transformation aftermath, and its development opportunities remain uncertain."[30]

At the same time, however, the rural share of the population remained stable at roughly 30 percent throughout the decade. The proportion nominally employed in the agrarian sector still hovered at around 18 percent,[31] although hidden unemployment levels may have been as high as 50–60 percent.[32] Barring a radical reversal in the state's liberal policy orientation, real agricultural employment seemed bound to contract further in the near future, bringing Latvia's figures closer to the much lower levels in advanced industrial countries. As Latvia began integrating into European and global markets and institutions, the traditional relationship between land, labor, and Latvianness was radically destabilized. With one-third of Latvians still dwelling in the countryside, rural development represented a crucial challenge in the pursuit of overall economic and social recovery. This challenge was at the same time a question of identity: would Latvians remain Latvian if they were no longer a "nation of farmers"?

The Ulmanis Days: Back to the Future

Throughout the first post-Soviet decade, many Latvians viewed the demise of agriculture not as an ineluctable result of modernization but as a national tragedy to be averted at all costs. As in the 1920s and 1930s, the agrarian nationalists of the 1990s envisioned agriculture as the engine of national development. "The greatest wealth of the Latvian nation, the basis for its survival, development, and flourishing is the Latvian land," declared an agricultural economist.[33] Under news headlines declaring, "The nation's foundation is in the countryside" and "Latvia needs strong farmers," agriculturalists, politicians, academics, and artists reproduced the trope of agricultural labor as the preserver of spiritual, moral, and physical health, of the "national mentality," indeed, of Latvians as an "ethnos."[34] Diverse commentators identified Latvia's rural way of life as the only defense of cultural uniqueness against the forces of global and European integration. "Our fields and forests," declared a leader of the Latvian Farmers' Federation, will guarantee "the Latvian way of life, which through eight centuries of foreign occupation allowed us to preserve our national identity. Thus we will preserve our tough and stubborn people, who have always been shaped by the Latvian peasant *viensēta*."[35] Jānis Purapuķe's proverbial phrase, "one's own little piece of land," remained a universal shorthand for that purportedly essential element of the "Latvian mentality": the yearning for private landed property.

Agrarian nationalists insisted that rural population levels must be maintained and even increased relative to the big cities, with their growing crime rates and large Russian-speaking populations. "The ideal situation would be if at least 30 percent of Latvians were rural dwellers, even though this conflicts with all economic considerations," maintained a local government official. "Some sacrifices must be made to preserve the nation's mentality and for the sake of our young people's health."[36] "A healthy society, a healthy state cannot develop within city walls," concurred a leader of the Farmers' Union. "Only if society, the state develops its rural life can it defend the nation's spiritual potential. . . . Everyone who can find something to produce and sell must have the opportunity to do so."[37] In the present era of global economic competition and integration, of course, a large, labor-intensive agricultural sector cannot be maintained without extensive state intervention, and this is precisely what agrarian nationalists demanded. Most Latvian farmers themselves, reports Tīsenkopfs, were "convinced that their duty was to create a farm and then to engage (a bit) in production. The duty of the state, the majority feels, is to provide economic support for (small-scale) production."[38]

The notion of the countryside as the bedrock of Latvianness was sharply challenged in 1994 when the liberal Latvia's Way Party introduced an amendment allowing foreign-owned firms to purchase rural land, a right previously restricted to Latvian citizens. Agrarians countered by championing the right of Latvians to earn a living by farming. The director of the State Land Service maintained that permitting legal entities to buy land without the intention of cultivating it would exacerbate rural poverty and unemployment.[39] A member of parliament warned that rural land sold to foreigners by cash-strapped Latvians "would largely be used for recreation, instead of economic activity," thereby forcing former farm owners into demeaning jobs as dishwashers and the like.[40] Just as for the advocates of "Wise Use" of the American West, described by James McCarthy, "the pivotal question was not so much *whether* they could make a living at all, as *how* they would make it."[41] What was at stake was not only economics but also the moral economy of the countryside. In post-Soviet Latvia, the interwar peasantist ideology had been reinforced by the Marxists' privileging of labor and material production that dominated the Soviet period, resulting in the widespread belief that all economic and moral values come from working the soil.[42] Denouncing "Europification" as the newest form of colonization, agrarians bitterly inveighed against the prospect of the Latvian countryside being used to entertain European tourists. The "little brother with the centuries-old dream of his own piece of the earth" was pitted against the "world citizen," the hardworking farmer against "speculators" and financiers.[43] The liberal camp eventually prevailed, though only in late 1996, after several years of acrid polemics.

At the heart of post-Soviet agrarian nationalist discourse was a vision of the interwar First Republic, especially the Ulmanis days, as a golden age and a developmental template for Latvia's second era of economic and political reconstruction. The Ulmanis days were widely remembered as a time of national pride and prosperity, when the nation lifted itself from the devastation of the war on the backs of its hardworking family farmers. Latvia's first post-Soviet president, Guntis Ulmanis, attained his position entirely by virtue of being the nephew of the self-designated Leader (*Vadonis*), remembered primarily not as a dictator but as the benevolent guardian of the nation. In the first post-Soviet parliamentary elections in 1993, the campaign posters of the resurrected Farmers' Union Party featured the countenance of the elder Ulmanis against a background of well-tended private farm fields, above the slogan "For Latvia—national, beautiful, and strong!"[44] In December 2002, the Riga City Council approved the building of a monument to Ulmanis across the street from city hall. The winning design, selected in a competition organized by the Farmers' Union Party, portrayed the leader wading through a thigh-deep field of rye in the shape of Latvia. (Protesting the conservative tenor of the official competition, Latvian art students staged an alternative exhibit, called "Ulmanis in Our Hearts," which featured representations of Ulmanis as "a figure in butter, a portrait on watering cans and a plastic statuette with a mobile phone on a marzipan cake." The alternative exhibit drew outrage, including a picket by the patriotic youth organization Everything for Latvia.)[45]

At a 1997 conference in honor of the 120th anniversary of Ulmanis's birth, scholars defended his agricultural support policies against liberal critics. The Ulmanis regime was lauded for devoting 17 percent of the national budget to agricultural subsidies and for promoting education in rural areas, unlike its parliamentary predecessors and unlike post-Soviet governments.[46] Conference participants sided with the agricultural economist Artūrs Boruks in eschewing the term *dirigisme* when speaking of the Ulmanis regime, preferring instead "organic planning-type system" or "regulated and state-led people's capitalism." According to Boruks, state intervention under Ulmanis sought to promote "capitalism and welfare not for a few large landowners, but for the whole peasantry."[47]

Similarly widespread was the notion that Soviet occupation had eroded the values and work ethic of the interwar era. In those days, according to Heinrihs Strods, "family farms were characterized by a strong work ethic, . . . self-discipline, and positive work traditions."[48] The multimillionaire and three-time prime minister Andris Šķēle lauded the Ulmanis days as a time when "every child was responsible for his parents, parents for their children and grandparents—the whole family hierarchy was ruthlessly enforced." In contrast to today's rampant corruption, claimed Šķēle, "Ulmanis was frugal with state money, and he kept state funds under lock and key. Ulmanis didn't flit about on foreign junkets, he didn't fritter away the state, but provided for it, saved."[49] "It will be in the countryside that the Latvian state will be reborn," wrote a pensioner in 1997. "I remember the first period of Latvian independence. People got rich only through work, and not through any eight-hour workdays. No one complained. . . . In the first Ulmanis days, people ate in order to live, but when I watch television these days, it seems like some people live in order to eat."[50]

The "Singapore of the Baltic"

The veneration of Ulmanis and the Ulmanis days was so widespread as to become a national cliché and, as such, a prime target for attack. In place of a "far-sighted and realistic agricultural policy," complained one critic, Latvia had only "nostalgic memories about the golden days of Ulmanis, when silver coins ceaselessly rolled into farmers' wallets through the export of butter and bacon."[51] A philosophy student lamented: "We are still immersed in the heavy romance with the land,

with the dark depths of the past." Another wondered how Latvians might over-
come their attachment to "peasantness" (baurība).[52] A sociologist expressed irri-
tation at the "theory of two types of Latvians—the real ones living in the countryside and those other ones," and argued that "we cannot simultaneously fear the
Russians, hate Europe, and plant up all of Latvia with potatoes."[53] A historian observed:

> Society has to a large degree returned to the ideology of the 1930s. The schools and
> mass media cultivate erroneous notions of Latvians as a peasant nation, of "700 years
> of slavery," of the primacy of folklore in the development of Latvian culture and so on.
> . . . The dominant historical myth about a peasant nation's battle with the black
> knight has long since lost any connection to social reality. The fact that "Latvian" is
> not a synonym for "peasant" was already established by the New Latvians in 1862.[54]

Some critics of the agrarian discourse self-consciously assumed the mantle of
Valdemārian antiagrarian internationalism, linking Latvia's developmental destiny to the forces of globalization and European integration. Invoking the familiar
trope of the bridge between East and West, commentators, business people, and
politicians pitched Latvia as "the Singapore, or Hong Kong, of the Baltic."[55] "We
are in a good position [to become] a bridge," declared a prominent banker. "Baltic banks speak three languages fluently, and this is very important in relations
with customers."[56] The liberal parties that dominated Latvia's post-Soviet cabi-
nets and parliaments welcomed foreign investment and ownership, facilitated the
creation of various duty-free economic zones, negotiated swift entry into the
WTO, and lobbied aggressively for Latvia's accession to the European Union. Like
Valdemārs, in short, internationalists in the 1990s looked not to the land but to
the sea. They saw the future in Latvia's geography of transit, in its position as a
gateway "between two divergent economic systems."[57] In light of global and regional market trends, they viewed agriculture as far less promising than international trade, tourism, financial services, and information technologies, and they
rejected the agrarian notion that "material production is the economy's foundation and the rest is merely parasitical."[58]

Liberal internationalists firmly advocated rural economic diversification, although they did not necessarily believe it could bring prosperity for a third of
Latvia's residents. Thanks to the rural-to-urban migration of the Soviet era, the
population density in Latvia's countryside was already extremely low, creating a
poor environment for development of industry, services, or even basic infrastructure.[59] Liberals envisioned the concentration of people and nonagricultural economic activities in provincial towns and saw no alternative to depopulation for

much of the rest of the countryside. According to Andris Miglavs, director of the Latvian State Institute of Agrarian Economics: "We must begin to recognize clearly that only the most successful farmers will live in farmsteads, while the rest will make their home in villages and towns." Areas lacking "marketable natural resources" or public services faced inevitable depopulation, he maintained, and "better that it be a planned depopulation, maybe with some rational concentration of residents in villages and towns. . . . From an economic perspective, I really see no other vision at the moment."[60] As for agricultural production, the state should concentrate its support on the limited number of already profitable farms. "If we 'loved' everyone who wants to pursue traditional agriculture," noted Aigars Štokenbergs, a rural development project manager at the World Bank, "we would sooner or later sink to the poverty typical of an agrarian economy."[61]

This line of reasoning was, of course, anathema to agrarian nationalists, who reviled the notion of Latvia as gatekeeper between East and West and of integration into European institutions and global trade as the lynchpin of development. "We have barely gotten out of one 'kolkhoz,'" went the common refrain in opposition to the EU, "and we are already trying to steer ourselves into a new one."[62] "The Maastricht treaty hangs like a sword of Damocles above the peoples of Europe. American big agribusiness threatens to flood Europe with cheap agricultural products. One feels that the countryside and along with it the nation's foundation will be destroyed."[63] Echoing the isolationism of the 1930s, agrarians insisted that the basis of the national economy should be domestic production, not "a cargo transfer facility on the shores of the Baltic Sea. . . . Latvia must remain Latvia and nothing else, a state with Latvian, not cosmopolitan features. Latvia in her citizens' consciousness must be a home, not a way-station, whose owner is destined merely to play in the sand along the roadside and greet the passers-through."[64] The survival of the Latvian nation, agrarians insisted, depended on keeping rural space filled with Latvian dwellers. As Arturs Boruks put it:

> Agriculture is not just one among many economic sectors. Agriculture fulfills other important, irreplaceable functions: it solves ecological problems, ensures preservation of a healthy environment. . . . But the most important, irreplaceable function is the filling up of the countryside with people—the preservation of our land for future generations, without which neither the nation nor the state can exist.[65]

In the late 1990s, Latvian society was still a very long way from abandoning what Boruks's liberal colleague Miglavs called the "myth that we must return to the interwar rural population structure, when all, or at least the overwhelming majority, of Latvia's rural dwellers lived in *viensētas*."[66]

Indeed, while the post-Soviet agrarian reform had spurred a large-scale return to the *viensēta* in the early part of the decade, the wrenching liberalization of the ensuing years had forced many farmers to abandon their new farms, or at least to shrink their production to small subsistence plots. "Since the regaining of independence," declared an outraged manifesto of the Latvian Rural Support Association, "political parties in Latvia, through wrongheaded and selfish activities, have allowed agricultural production to shrink by 60 percent; more than 400,000 hectares of agricultural land have gone out of production."[67] Agrarians mourned the transformation of the post-Soviet countryside into a landscape of abandonment: a "ghost landscape," in Edmunds Bunkše's phrase,[68] of fields grown over with the brush bitterly referred to as Latvian cotton, dotted with rotting barns and the picked-over skeletons of collective farm buildings. Abandoned land struck directly at the heart of Latvian agrarian nationalism, for it represented the abandonment both of the productive labor seen as constituting the genuine Latvian citizen, and of the solitary farmstead that reified the laboring relationship between land and self.[69] As Anda Līce bluntly put it, "The devastation of the countryside is endangering the spiritual survival of our people."[70] According to Miglavs, it had been a key slogan of the glasnost-era independence movement that "one of the chief harms wrought by the Soviet regime was the fact that a million hectares of land had gone out of production" due to collectivization. Ten years later, agrarian nationalists wielded the same slogan against the liberal policies of postindependence Latvian governments.

Land abandonment and rural depopulation in agriculturally marginal areas are feared in many West European countries, too, but these are only potential threats in most EU member states, with their tremendously high population densities. There, the key challenge is balancing competing demands on scarce rural land. In Latvia, by contrast, the emptying of the rural landscape was symptomatic of land surplus: outside of the fertile central regions, most agricultural land had little or no market value, and some 250,000 hectares had no owner or claimant. To keep people gainfully employed in the countryside, Latvia faced the quite different challenge of creating economic value in a devalued rural space.

As in Valdemārs's time, in the 1990s the most prominent symbols of the post-Soviet internationalist discourse of identity, homeland, and developmental destiny were Latvia's multicultural cities and ports. Riga had reclaimed its place among the cosmopolitan capitals of Europe, with its breathtakingly beautiful restored Jugendstil facades, luxury import boutiques, German tour buses, and armies of cellular phone–wielding business people, foreign diplomats, and young polyglot Latvian professionals. The free port of Ventspils on the northern Kurzeme coast, home to a booming transshipment facility for oil piped in from Russia, was Latvia's most notorious cash cow throughout the 1990s.[71] Aivars Lembergs, the long-serving Ventspils mayor and transit trade "oligarch," urged his countrymen "not to think that Latvia begins with a hog's nose and ends with a hog's tail."[72] In the post-Soviet era, however, not only the cities and the ports but even the countryside—historically a haven of relative ethnic homogeneity and cultural isolation—was being reimagined in terms of a geography of transit and openness.

In 1991, a majority of Latvians, influenced by the potent national myth of the agrarian golden age of the twenties and thirties, had expected Latvia's economic development to be fueled by agriculture. It did not, perhaps, require a very great leap of faith for the well-off household plot holder to become a private family farmer, and yet the salaried kolkhoz employee was worlds away from the vicissitudes of globalization and "market discipline." In the 1990s, Latvia's liberal governments were unwilling or unable to protect farmers against those vicissitudes. As successive governments remained committed to trade liberalization and accession to the EU, it became increasingly evident that the number of Latvia's rural dwellers deriving gainful employment from agricultural production could only decline in the coming years and that abandonment of Latvia's already devalued rural land would accelerate.

In the agrarian discourse of Latvianness, as we have seen, farming, nature, and national identity are tightly interwoven: both nature and the Latvian character are "produced" through agrarian labor. The marginalization of farming threatened both sides of this equation. Post-Soviet economic liberalization was finally forcing Latvians into a fundamental renegotiation, paradoxically postponed by fifty years of Soviet rule, of the relationship between land and Latvianness. For the first time since the peasantist turn of the 1920s, the hegemony of agrarian notions of nature, homeland, and developmental destiny was being seriously threatened. Contemporary globalization seemed to be succeeding, where Sovietization had failed, in disengaging Latvianness from landed labor in a historically unprecedented way.

Part 2
Contesting Nature after Communism

Chapter 4

Returning to a Postproductivist Europe

Liberal internationalists, as we have seen, accepted the disengagement of Latvian-ness from landed labor as an inevitable by-product of modernization, although not without some reservations. The economist Andris Miglavs conceded that a "healthy society and state" could not be based in a "solely urbanized environment," but had to be "rationally, sustainably rooted in the whole land, the whole environment. And one of the elements of that environment undeniably is the farm, and one of the elements of the farm is agricultural production."[1] During both the interwar and Soviet periods, the reigning productivist values ensured that governments would act decisively against the emptying of the countryside by subsidizing farms and importing foreign agricultural laborers. In the 1990s, agrarian nationalists made similar arguments for subsidies and market protection, but in the context of Latvia's race to join the WTO and the EU, this strategy was no longer viable.

In its place, some liberal internationalists were embracing a new "postproductivist paradigm" that was gaining ground in the EU. Overproduction and market liberalization, along with the powerful global narratives of sustainable development and biodiversity, was fostering a fundamental reassessment of rural land. Some Latvians began to see the future of the countryside and its residents not in exporting butter and bacon, as in the Ulmanis days, but in "exporting" wilderness and biological diversity to nature-starved West Europeans. Part 2 turns to this reimagining—and resistance to the reimagining—of the rural ethnoscape.

Greening the Common Agricultural Policy

With a communiqué of the European Council in June 1993, the EU signaled its formal readiness to begin "returning" Eastern Europe to the fold. The Phare program, launched in 1989, had by 1992 been expanded to include the other Central and Eastern European countries (CEECs).[2] Following the regional pattern, Latvia signed a Free Trade Agreement with the EU in January 1995, became an associate member by signing a Europe Agreement in June 1995, and opened formal negotiations for accession in 1998. It was officially accepted as a candidate in October 1999, along with seven other CEECs, with 2004 the target date for accession.[3] Latvia, like the other new candidates, was much poorer than even the poorest member states and had a much larger and less productive agricultural sector.[4] While the post-Communist countries struggled to restructure their agrarian economies, the EU faced the equally daunting challenge of integrating them into its already strained institutional framework.

In particular, eastward enlargement exacerbated the already urgent need to reform the EU's system of agricultural supports, the Common Agricultural Policy (CAP).[5] By many accounts, the defining weakness of the CAP was the policy of "coupled" price supports, which tied income support for farmers to output of particular agricultural commodities.[6] Coupling had become increasingly indefensible as the EU sought to liberalize trade in negotiations with the General Agreement on Tariffs and Trade (GATT) and its successor, the WTO. At the same time, the environmental impacts of the CAP had come under increasing scrutiny. Armed with the global narratives of biodiversity and sustainability, European environmentalists articulated a compelling indictment of CAP-supported agriculture. Coupling rural income assistance to output, they demonstrated, had promoted intensification—that is, boosting land productivity through increased use of chemicals and fertilizers, allowing higher ratios of livestock to land, wetlands drainage, river realignment, converting marginal land into production, and monoculture planting. These practices not only caused pollution but also destroyed and fragmented natural habitat. On the other hand, conservationists showed that traditional, nonintensive farming practices supported biodiversity, noting that most of Europe's present landscape had been shaped by centuries of human management. Jim Dixon writes:

> Europe has very little true wilderness left, except in the far north and the tops of mountains. Nearly all its animals and plants rely on "semi-natural" habitats which have evolved over several thousand years of agricultural exploitation. Where systems of farming have remained unchanged for many centuries, very rich or specialized plant and animal communities have developed and these represent some of the principal biodiversity resources in Europe.[7]

In other words, if capital-intensive farming simplified ecosystems and eliminated specialized habitat niches, then low-intensity farming—known also as low-input, extensive, or high nature-value farming—served precisely the opposite function.[8] Many European conservationist organizations thus began turning their attention to protecting biodiversity not only in nature preserves and other protected areas, but also on farmland. This new conservationist agenda was reinforced by a shift in popular consciousness. As agricultural productivity and surpluses continued to grow, and as Europe became increasingly industrial and affluent,[9] well-off city dwellers, freed from the food security concerns of the lean postwar years, took up the cause of preserving Europe's traditional agrarian landscapes.

In 1992, the EU formally committed itself to sustainability in its Fifth Environmental Action Program. Beginning in the mid-1980s, a range of agri-environmental measures were implemented to promote extensive farming practices, particularly in seminatural grasslands and other so-called high nature-value farming systems.[10] Nonetheless, many observers claimed that while some gains had been achieved directly or indirectly in environmental and landscape protection, these measures had been gravely underfunded and outweighed by the CAP's fundamental promotion of intensification. Moreover, the sustainability narrative sharpened criticism of not only the CAP's impact on biodiversity, but also its failures in promoting rural development, broadly conceived. In an era of growing production surpluses, critics of the CAP began to argue that "farmers and land managers can and should at the same time supply other goods and services possibly inherent in rural areas," including "afforestation; rural sites for non-agricultural development . . . leisure and recreation, and of course, the preservation of habitats, wildlife and the rural cultural landscape in all its forms."[11] By the late 1990s, the buzzwords of the new paradigm had become ubiquitous within bureaucratic and nongovernmental circles alike: diversification of the rural economy; a multisectoral, integrated approach to rural development; rural development as the "second pillar" of the CAP, along with production support.[12] This approach was formally endorsed in *Agenda 2000*, the strategic planning document agreed upon by member states at the Berlin Summit in March 1999. The document stated, "Agriculture must play its role in preserving the countryside and natural open spaces, and . . . must make a key contribution to the vitality of rural life."[13]

From the mid-1980s onward, in short, overproduction had converged with the powerful discourses of free trade and sustainability, and with the shifting values of domestic electorates, to launch a postproductivist transition in European rural policymaking. The European countryside was being reimagined as a site no longer primarily for agricultural production, but for providing environmental, aesthetic, and recreational amenities, and farmers were being reinvented as "stewards of the countryside" and "guardians of the landscape." Policymakers and urban dwellers had come to view the countryside as a provider of more than agricultural output, and conservationists were increasingly focusing on farming practices as crucial agents of both biodiversity loss and biodiversity conservation. Given the hegemony of the free-trade doctrine, the eventual decoupling of rural income support from output seemed inevitable, and conservationists were taking advantage of this fact to lobby for "green recoupling": that is, a policy of subsidizing farmers directly and explicitly for protecting the environment and the cultural landscape.

While the future shape of European rural policies in the neoliberal era remained uncertain, optimists maintained that rallying political support for these increasingly valued public goods would be more feasible than justifying continued income support for farmers per se, particularly since the CAP now consumed nearly half of the total EU budget. Skeptics predicted, however, that mustering sufficient financial and political resources for a genuine green recoupling would be an uphill battle. Without effective policies to promote low-impact farming, they warned, ending price supports might instead mean that production was concentrated in fewer and larger intensive farms, driving small, traditional farmers out of business. Biodiversity would then suffer both from concentration and from land abandonment. And land abandonment would change the social face of the countryside, clashing with the hallowed conviction of many Europeans that "an economically viable and beautiful countryside is one managed by large numbers of farmers and that a principal concern of rural policy should be to keep farmers on the land."[14]

Public debates about the contours of Europe's postproductivist transition were structured by the narratives of biodiversity and sustainability, but as noted earlier, these narratives are profoundly ambiguous and hotly contested. The seminal Brundtland Report (1987) acknowledged that "there is no single blueprint of sustainable development, given that economic and social systems and ecological conditions differ widely among countries."[15] As late as 1997, "even a quick glance at the large volume of writings on sustainable development reveals that there is no general agreement on exactly what sustainable development means."[16] Biodiversity is a similarly broad concept and resists precise definition. Scientists cannot agree on what constitutes a species, much less on the parameters of something that can be defined as broadly as "life and all that sustains life."[17] The scope for varied and conflicting interpretations is even greater when determining which elements of biodiversity are most valuable and devising strategies for their protection or sustainable use. The postproductivist paradigm, in short, was not accompanied by ready-made "toolkits" or easily translatable "blueprints" for implementation.

Importing Sustainability

This ambiguity did not prevent the EU from exporting the new paradigm to its eastern aspirants. From the earliest years of the post-Communist transition, Latvia, like the other CEECs, was awash with foreign environmental assistance, primarily from European donors as "part of a deliberate strategy for the Europeanization of Eastern Europe and development of what was referred to as a 'Common European House.'"[18] Much of this aid was devoted to reducing crossborder pollution—in the Latvian case, primarily pollution of the Baltic Sea.[19] But sustainable development and conservation of biodiversity have also been targeted by a host of bilateral, intergovernmental, and nongovernmental donors, including Phare, UNDP, the Global Environmental Facility (GEF) administered by the UN and the World Bank, WWF-International, Birdlife International, and the governments of Denmark, Finland, Germany, the Netherlands, Norway, and Sweden.

The Latvian Ministry of Environmental Protection and Regional Development quickly became a hotbed of international involvement. Its staff was trained in Western methods, participated in international study tours, and worked side by side with resident Western consultants. Latvian environmental officials were quick to adopt the new language, declaring as their top priority harmonizing Latvian legislation with global and European norms. A host of international treaties on sustainability and biodiversity were swiftly ratified. Among the most important were the Ramsar Convention on Wetlands of International Importance Especially as Waterfowl Habitat, the Convention on International Trade with Endangered Species of Wild Fauna and Flora (CITES), the Bern Convention on the Conservation of European Wildlife and Natural Habitats, and the Rio Convention on Biological Diversity. To comply with international and European norms, the ministry produced a range of strategic documents and action plans acknowledging the importance of sustainability and biodiversity. The National Environmental Policy Plan of 1995, for example, declared that "a quality environment is a prerequisite for sustainable development in Latvia,"[20] and the 1998 National Report on Biological Diversity affirmed that "biological diversity is one of Latvia's most significant treasures."[21]

The more conservative agrarian sector was slower in responding to the new paradigm. In sharp contrast to the heavily internationalized environment ministry, as late as 1996 the Ministry of Agriculture possessed only two computers, used solely for word processing, with no e-mail or Internet access. Agricultural policy remained staunchly productivist and focused nearly exclusively on promoting conventional farming.[22] In 1997, however, a two-year Phare project was launched to restructure the ministry and strengthen institutional capacity in preparation for accession to the EU and participation in the CAP. In addition to providing several foreign experts, the project trained a dozen local junior experts in various aspects of EU policy for agricultural and rural development. They set to work analyzing and translating the relevant portions of the *acquis communautaire*— the voluminous body of EU norms and regulations with which all aspiring members are required to comply—and organized committees to hammer out proposals for harmonizing Latvia's laws with them.

In 1998 Latvia received a World Bank credit line, supported by various Phare pilot projects, to promote "diversification of the rural economy," or nonagricultural forms of generating rural income. According to Andris Miglavs, diversification "as a slogan" played a critical role in beginning to dispel the potent national "myth . . . that rural dwellers' work has to be tied to agriculture."[23] In June 1998, the parliament approved a Rural Development Program with the stated goal of identifying the preconditions for integrated, diversified, and sustainable rural development. With this document, "the fact that agriculture and the countryside are not equivalent, but only related terms, [had] been politically recognized."[24] By the end of the decade, in other words, Latvia had officially acknowledged the need for promoting sustainability in the rural sector. While the buzzwords of postproductivism had been widely adopted, however, there was considerable variation in depth of commitment to the new paradigm and uncertainty as to what sustainable rural development might actually look like in practice or how to achieve it.

Latvia's Europe Agreement went into effect fully in February 1998, establishing the formal framework for beginning entry negotiations and receiving preaccession technical assistance.[25] An interministerial task force was charged with drafting a Rural Development Plan specifying what measures Latvia would undertake to prepare for accession to the EU. Latvia was the first applicant country to submit a draft, prepared at breakneck speed in 1999. The draft plan identified five target areas for needed EU subsidies, including diversification of the rural economy and environmental measures pertaining to agriculture. The accompanying text acknowledged the importance of protecting both biodiversity and cultural landscapes:

> Through the centuries, a unique and diverse landscape, which preserves the nation's historical and cultural heritage, has developed in the territory of Latvia. During the last fifty years some of this territory has suffered a reduction of biological diversity and the intentional or unintentional destruction of landscape. . . . The intensification of economic activity poses serious threats to Latvia's still relatively rich nature and landscape; thus it is imperative to ensure protection of landscape elements. The mo-

saic landscape, and particularly natural meadows and grasslands, are crucial to the preservation of biological diversity.[26]

Yet both EU and Latvian analysts argued that the draft plan was biased heavily toward modernizing the conventional farm sector, reflecting the interests of a small but influential elite who owned large and growing farms. The plan had been drafted too quickly, they maintained, with no time for genuine consultation with stakeholders beyond the central staff of the agriculture ministry. It was far from clear whether the language of the plan reflected any real political support for economic diversification and environmental protection, and whether state or private cofinancing for these measures (without which EU subsidies could not be received) would realistically be obtainable.

Moreover, responsibility for rural policy in Latvia was parceled out among several ministries and lacked an institutional center.[27] The new plan failed to integrate the proposed initiatives with ongoing rural development programs administered by agencies in other sectors, and the government appeared to lack the financial and human resources needed to bring about such integration. Only a handful of overworked staff members (supplemented by one Phare-sponsored junior expert) at the agriculture and environment ministries were assigned to work on policies for environmental protection and sustainable development. Most fundamentally, according to many observers, the draft plan was not based on any genuine vision for rural development, nor did it seem a workable policy tool. Rather, it was merely a mechanical response to EU requirements, a formal "paper for Brussels."[28]

Despite the proliferation of required documents such as the Rural Development Plan and the Rural Development Program, by 1998 Latvian analysts concluded that "Latvia still lacks a genuine structural, environmental, and social policy directly targeted at the problems of rural development."[29] While formally acknowledging the need to promote alternative, nonagricultural uses of rural land and forms of rural employment, the state had not yet articulated a substantive vision of what these alternatives should be or how to begin realizing them. Agricultural officials appeared to be clinging to the productivist paradigm, paying only lip service to foreign advisers' warnings about the eventual "greening" of EU rural aid schemes. They had mastered the art of Western "project-speak," it seemed, without internalizing Western norms and agendas.[30]

In the interwar period, rural development had been an overriding political priority for the new Latvian state, which acted decisively to reshape Latvians' relationship to the land. In the 1990s, however, the new Latvian state lacked both the

administrative capacity and the political will to decisively guide the postproductivist transition. The majority of the electorate were urbanites, after all, and radi-
calization of the disaffected countryside seemed unlikely, despite recurring minor
protests against subsidized sugar beets from Lithuania or hogs from Estonia.
Without definitive action by the state, the shape of Latvia's postproductivist transition remained indeterminate and ripe for contestation. In this moment of radical uncertainty about the future, the dual discourses of liberal internationalism
and agrarian nationalism provided the conceptual maps with which Latvians began to imagine how rural land should be used in a postproductivist era.

Capitalizing on Nature

With the first visits of German bird-watching enthusiasts in the late 1980s, Latvians began to realize that in their landscape's biodiversity, they had something to
offer the West. As an environmental official observed in 1991: "Half of the species
identified in the Bonn Convention [on the Conservation of Migratory Species of
Wild Animals] are common in Latvia, while in many European nations they can
now be found only on postage stamps."[31] To a large extent, this wealth of wildlife
was a fortuitous legacy of Russian domination: the relative economic inefficiency
of imperial Russian and Soviet socialist rule meant later and less intensive development of agriculture, logging, and industry in Latvia than in the West. Addition-
ally, the Soviet imperative to secure the western border militarily preserved large
tracts of relatively untouched nature, especially along the 500 kilometers of Baltic
Sea coast, which is an important flyway for many migratory bird species.

Latvia's abundant wetlands are also critical resources for biodiversity. Thanks
to land abandonment resulting from collectivization, large areas of wetlands had
survived Brezhnev's land reclamation campaign.[32] These areas now support ro-
bust populations of many wetland species that are threatened elsewhere in Europe, including lesser spotted eagles and as much as 10 percent of the world's
population of black storks.[33] Latvia's wet forests are an especially rich habitat resource. A million hectares of abandoned agricultural land became swamp forest
during Soviet times, and today over 20 percent of Latvian forests are on wet soils,
providing refuge for not only eagles and storks but also wolves, lynx, beavers, otters, and other threatened species.[34] According to the World Conservation Union
(IUCN), "Such areas could today have a major conservation role, because of the
wholesale destruction of forests on wet soils" in Central and Eastern Europe.[35]
"We have visited wetlands in many countries around the world, but never have we
had the opportunity to see such a natural bog as here," declared the London-based

director of an international wetlands research organization when visiting an important wetland preserve. The bog, he said, was "like an open textbook for us."[36]

In Latvia as throughout Central and Eastern Europe, the post-Soviet economic collapse was in many ways a boon for nature, as poverty prevented many farmers from maintaining or expanding drainage systems and from using chemicals, fertilizers, and other weapons of intensification. The mass restitution of land, however, had placed the future of Latvia's biodiversity resources in the hands of hundreds of thousands of private owners who could quickly decimate them through intensive farming, logging, or construction. "Biodiversity is one of Latvia's undervalued riches," remarked Valts Vilnītis, environmental adviser to Latvia's first post-Soviet cabinet, "a vast capital that could easily be squandered during the period of economic transition in the race for immediate affluence, but which is extremely difficult to renew even in a wealthy nation."[37] How best to steward—and perhaps profit from—this "vast capital" in the post-Soviet, postproductivist era? This was the challenge facing Latvian nature management professionals in the 1990s, and their responses were structured by the tension between agrarian and internationalist discourses.

Figure 5. A wealth of biodiversity. Photo: Māra V. Kore.

As we have seen, Europe's postproductivist paradigm had ambivalent implications for agriculture: while it devalued agriculture as the primary function of rural land, it also created the new category of "high nature-value farming," which upheld traditional, extensive farming methods as crucial tools for conserving biodiversity and thereby reinforced the cultural argument for maintaining a peopled agrarian landscape. In Latvia as in Western Europe, current biodiversity resources reflect centuries-old patterns of human use. Not only undrained wetlands, but also meadows, pastures, and cultivated agricultural lands are rich sources of wildlife habitat. The lesser spotted eagle nests at the forest's edge and feeds in fields. Mating pairs of white storks return year after year to nests atop old chimneys or telephone poles—as potent a symbol of the Latvian ethnoscape as the great tree (dižkoks). In Latvia as in the EU, agrarian nationalists could invoke biodiversity to argue against the abandonment and conversion of agricultural land.

A prominent exponent of this view was Jānis Priednieks, then director of the nongovernmental, quasi-academic Latvian Fund for Nature (Latvijas Dabas Fonds, or LDF). Founded in 1990 by biologists from Latvian academic and research institutions and supported by project funding from the Latvian government, the EU, and West European governments, the LDF is Latvia's leading organization for preparing and implementing nature management plans for protected areas. For Priednieks, the greatest potential threat to biodiversity in Latvia was not the future intensification of agriculture, but rather the diminution of total agricultural land in production. While acknowledging that "Latvia, as a modern Western state, will not be able to afford preserving all of its farmland if [agricultural] exports to Russia shrink in the future," he was nonetheless optimistic about Latvia's prospects for breaking into the growing market for organic produce, as well as receiving eventual EU support for low-impact farming. "It remains to be seen what shape the European Union's agricultural reform will ultimately take," he conceded, "but we are very naively hoping that if Latvia gets money, it will be for nature-friendly, and not for intensive, agriculture."

Priednieks defined Latvia's biodiversity primarily in terms of the species associated with agricultural landscapes, rather than, say, undrained wet forests. From this perspective, the key to conservation was to secure EU subsidies for traditional farming, and especially dairy farming, to prevent the seminatural grasslands, arguably Latvia's most ecologically valuable agrarian ecosystems, from growing over. He opposed using EU subsidies for afforestation of marginal farmland, and while he welcomed ecotourism, he saw it as entirely dependent on farming: "The grazing of dairy livestock is the foundation of Latvia's rural diversity, and I think also of rural tourism to a large degree."[38] Unlike many unreconstructed protectionists in the agricultural sector, Priednieks was not hostile to the international flow of ideas and market forces in general. Indeed, the LDF collaborates with international NGOs and foreign governments on a range of projects and receives much of its funding from abroad. Rather, he took advantage of the EU's own lack of consensus on what a sustainable, postproductivist countryside should look like and what exactly biodiversity is and how to protect it in order to defend his understanding of nature as agrarian ethnoscape.

In Latvia as in the EU, the conservation-based argument for farming mutually reinforced the celebration of nature as a cultivated agrarian landscape and the act of cultivation as central to national identity. Maruta Kaminska, an ecosystem inspector with the Liepāja regional environmental agency, shared her concerns about the future:

> I can't believe that Latvia will turn into one big fallow, for we are, after all, a nation of farmers [zemnieku tauta]. Except perhaps if we enter the EU—because it's not in their interest to promote our production. But if Latvia becomes one big fallow, then I think Latvia will disappear altogether, for what will we do then? . . . Either Latvia will be connected to agriculture, or else it will be the end.[39]

At a meeting of a task force on EU policy for agriculture and forestry, a participant insisted: "If we have good land, drained land, then to afforest it would be unpatriotic!"[40] Latvia's official environmental rhetoric affirmed both the global meaning of nature as a source of biodiversity and its national meaning as a reservoir of culture and identity. The National Environmental Policy Plan and the National Report on Biological Diversity both invoked the primordial Latvian closeness to nature, rooted in pre-Christian animistic religion and articulated in folk songs:

> For many generations most Latvians have lived in close harmony with nature, a lifestyle which has its roots in pre-Christian times when the people believed in natural deities. Elements and expressions of this attitude towards nature, full of respect and love, can be found in the Latvian Dainas—a unique historical collection of folk verse that bears witness to the peoples' [sic] way of thinking and perceiving. It is characteristic in the Dainas that even predatory animals such as wolves and crows are referred to in the diminutive form. It is as a result of these pagan attitudes that Latvia is richly endowed with "noble trees" [dižkoki], and rivers, lakes, springs, woods, caves, stones cliffs, etc. . . . are also sacred and protected.[41]

Latvia's natural heritage had to be defended not only to satisfy international norms and to capitalize on international market demand; it was also the embodiment of the "national consciousness and culture of the nation."[42]

Globalizing the Ethnoscape

If abandoned and devalued farmland was both a national and an ecological tragedy for agrarian nationalists, then for liberal internationalists it represented an opportunity to detach the value of land from farming and link it instead to global norms and markets—including markets for nature as a reservoir of biodiversity. While West Europeans burdened with their own production surpluses were not eager to import traditional agricultural goods from Latvia, much less to subsidize production there, they were potentially much more interested in "importing" Latvian nature via the growing business of ecotourism—tourism based on observing nature or experiencing the rustic simplicity of traditional farm life. "Tourism can support Latvia's economy more than agriculture," insisted environmental official Valts Vilnītis:

> What will we show tourists? The [eighteenth-century] Rundāle castle? They have plenty of castles. . . . But the Horn of Kolka [in northern Kurzeme] is the best observation spot during spring migration for birds of prey in all of Europe. Nowhere else in Europe are there such long stretches of undeveloped beaches. Many Europeans would pay real money to walk along a beach that doesn't have three German sunbathers per meter![43]

"Everyone is constantly talking about how we must enter Europe," observed a retired biologist and new landowner who chose not to farm his ancestral homestead but rather to capitalize on its marshes and "virtually primeval" forest by developing a nature park. "Nature is what we can still trump them with. We want to prove that by managing the land on the principle of environmental quality, in the most ordinary rural province and the most ordinary farmstead, birdsong, the peace and quiet of nature, the murmur of waters and forests can become market commodities."[44] The constraints of EU accession had spurred even the conservative agricultural establishment to endorse, on paper at least, this assessment of market trends. The draft Rural Development Plan stated, "The Latvian environment is highly suitable for the development of ecotourism, thanks to the diversity of forest ecosystems and the high occurrence of rare and protected species and habitats,"[45] and "Rural tourism is one of the most promising sectors of alternative rural enterprise."[46]

In an advertisement in London's *Financial Times*, Latvian tourist agencies sought to entice European visitors with classic Valdemārian imagery: "Located on the Baltic Sea, in the geographic center of Europe and a natural crossroads for trading and transport, Latvia and its capital city Rīga have assumed many traditions in the last centuries from other cultures." To this familiar trope they added the new language of ecotourism:

> Depending on the season, you can ski, fish, hike or pick mushrooms and berries. It is possible to view wild animals that have become extinct in western Europe. Fantastic landscapes will open up to you in the hilly highlands or ancient river beds, and seascapes in the golden sands of the seashore. A stay in a country farm house with its environmentally clean and natural lifestyle can prove to be very relaxing for the city dweller.[47]

Not only Latvia's seaports, but now also Latvia's nature and countryside were being reimagined as a landscape of transit and hybridity, penetrated by international values and markets.

This reimagining was facilitated not only by the hazy prospect of future tourism, but also more immediately by Western aid initiatives. In the mid-1990s, for example, the government of Finland funded an ecotourism development plan for the northern Kurzeme coastal zone, Latvia's most sparsely populated region and one of its most economically moribund. The Latvian-side project manager was the young geographer Andris Junkurs, who, much like Krišjānis Valdemārs, hoped to reclaim a forgotten Latvian seafaring legacy. The seventh through the ninth centuries A.D., he argued, were an unjustly forgotten period prior to German colonization, when Kurland's coastal towns "played as important a role as Riga during the Hanseatic era."[48] Finland had a special interest in the region as the home of the last concentrated community of Livs, a Finno-Uguric speaking ethnic group now largely assimilated into the Latvian mainstream. But "the German or the Swede also wants to know the links between this place and his own land," Junkurs added, "and then he can even feel more or less at home." While the task of agrarian nationalist homeland studies, from Atis Kronvalds through Imants Ziedonis, had been to make the Latvian feel at home in his own land, the vision of internationalist ecotourism endeavored to make the European feel at home in Latvia—in Junkurs's case, by forging what he called a Vikingesque environment from the fragments of an undervalued seafaring heritage. A geographical reimagining was needed to reinvigorate the marginalized coastal provinces. Echoing Valdemārs's claim that "Kurland and Livland are in the very center of Europe," Junkurs exhorted his fellow Kurlanders to think of the coast no longer as Latvia's remote border zone or Russia's western fringe, but as the center of the Baltic Sea region.[49]

In the southwestern corner of Latvia, nine rural townships banded together in 1993 to coordinate regional development planning by forming an association called Bārtava, after the Bārta River, whose watershed their territories encompassed. Learning from the European sustainable development paradigm, Bārtava members were beginning to see new value in their river valleys, groves, and marshes. "We have three 'aces,'" said Bārtava's director Gunta Strēle, a geography teacher and doctoral student in environmental education. "We are a coastal zone, a river watershed, and a border zone"—all "very modern" labels.[50] With assistance from Sweden and Norway, a Bārta River diversity utilization plan was being drafted. "You can't manage a river in just one segment," noted Strēle, "you have to take the whole watershed." Opportunities for EU funding had expanded area residents' perspective from the local to the regional and indeed to the international scale. Bārtava was working with three Lithuanian townships to establish a "Euroregion" that would qualify for EU cross-border development grants.[51] The new understanding of place as determined by ecosystems and Euroregions, like Junkurs's "Vikingesque" recentering of Latvian geography, represented a dramatic shift from the intensely local, inward-looking perspective of traditional Latvian Heimat protection, with its focus on the solitary and uniquely Latvian farmstead (viensēta) and great tree (dižkoks).

Arguably the most important role in transforming the Latvian understanding of geography and nature has been played by the World Wide Fund for Nature (WWF-International), the largest and one of the most influential nongovernmental actors in global environmental politics. In the forty years since its founding, WWF has evolved into a network spanning five continents and comprising twenty-six national organizations, five associates, twenty-two program offices, and almost five million individual supporters.[52] WWF began its involvement in Latvia by funding the preparation of a national conservation plan in 1991–1992. For WWF, Latvia was important insofar as it could contribute to conservation of continental biodiversity. Magnus Sylven, director of its Europe–Middle East Program, made this clear in his foreword to the conservation plan:

> Although Latvia is among the smaller countries in Europe, it is one of the richest in biological terms. Nowhere else in Europe does such a high density of otters patrol the banks of rivers and streams as in Latvia. More wolves are roaming the Latvian forests than in the whole of North-Western Europe, including Scandinavia. The possibility of encountering the secret black stork in a forest is not comparable anywhere else in Europe. Entering a Latvian wet forest in summer with the mosquitoes swarming around, the luxuriant ferns, the lilies, mosses and lichens, give one the feeling of being in the tropics. And the long, sandy beaches of the Baltic Sea are adorned with amber. Where else in Europe can you find almost 300 km of such beaches without hotels and other tourist facilities?[53]

WWF established a Latvian Program Office in 1993, staffed by Latvians and (until 2002) supported entirely through project funding from international donors, primarily West European governments and WWF national organizations.[54] Since its founding, the office has been directed by Uģis Rotbergs, who had initiated and directed the conservation plan project. As a young forestry specialist in the early 1980s, Rotbergs was already seeking out current Western research in forest science, ecology, and conservation biology. Upon joining WWF, Rotbergs came into more direct contact with Western ideas, participating in countless international meetings, conferences, and study tours in Europe and North America, and working side by side with European conservationists in implementing projects. Over the years, Rotbergs and his small staff have articulated a vision for conservation and sustainable development that departs radically from the Latvian mainstream.

This vision was informed by a recent paradigm shift in theoretical ecology and conservation biology. Whereas the so-called classical paradigm, developed in the early twentieth century, viewed natural systems as fundamentally closed and stable, since the 1970s the New Ecology has reimagined natural systems as open and dynamic. In the new view, nature is not a harmoniously balanced "web" tending toward "climax," but a "shifting mosaic," buffeted by natural disturbances like fire, drought, and animal grazing. Nature management should not aim to preserve a particular species or habitat, but rather allow the fullest possible functioning of natural processes within and between ecosystems.[55] WWF staff drew on the New Ecology to mount a sharp critique of mainstream conservationists like Jānis Priednieks of the LDF, who advocated preserving current farmland acreage to protect those species of wildlife that thrive in the agricultural landscape. Protecting these particular species should not be taken as the paramount goal of conservation, WWF argued, without first conducting a national discussion on what kind of nature Latvians want to protect or restore. Such a discussion had never taken place, lamented Rotbergs, even on such likely occasions as the drafting of Latvia's National Biodiversity Strategy. Instead, conservation specialists in government agencies and at NGOs such as the LDF or the Latvian Ornithological Association clung to the farm landscape as the a priori standard against which biodiversity losses and gains should be measured.

According to Mārtiņš Rēķis, WWF rural development project director, Priednieks's view of farmland as a uniquely valuable biodiversity asset holds up

> only if you take as the point of reference Latvian agriculture of the thirties, which ensured, what was it, 26 percent forests, something like 68 percent farmland, 10 per-

cent inhabited areas, and so on [as compared to today's ratio of 44 percent forests to 39 percent farmland]. . . . But I don't think that that's the ideal situation. I think that the ideal situation is one where as much as possible of this landscape is left the way it would be without human influence. Open spaces would still develop, because you would never get a solid forest cover. Of course, the lesser spotted eagle population would diminish, which is the most important thing to some people. And some kind of open, inhabited landscape would diminish too—but that, too, is not natural for Latvia; it came in together with agriculture, let's say 3,000 years ago. Or more precisely, people were there the whole time, they lived in swamps, clearings. . . . But this [human] population is inflated. . . . [Priednieks] thinks that this seminatural agricultural landscape is important. And I think that the natural landscape, such as it is, is important. Because, first of all, it's the least expensive to maintain. Because with the natural landscape you don't have to do much, it simply is there. And if you don't like it, you set a fire, burn a section. Then it will be open for a while—interesting![56]

Rēķis and Rotbergs contended that the sovereign goal of conservation in Latvia should be not to preserve the agrarian ethnoscape cultivated by the industrious farmstead owner, but rather to restore the preagricultural landscape with its shifting mosaic of open land and forests, continuously reshaped through the natural disturbances of fire, wind, grazing, predation, evolution, and so on.

Globally, WWF-International has embraced the political and economic principles of neoliberalism, seeking not to reverse but to harness for ecological benefit the forces of market liberalization, privatization, and capitalist development.[57] In Latvia, too, WWF accepted as inevitable the market-driven dwindling of Latvian agriculture. Whereas Priednieks based his agenda for biodiversity conservation on receiving EU subsidies for agricultural extensification, WWF argued that Latvia should prepare itself for the likely possibility that a thorough and adequate green recoupling (paying farmers directly for environmental protection) of EU rural support would not materialize. Subsidizing small-scale traditional farming was a costly approach to conservation, Rotbergs and Rēķis pointed out, and not sustainable in the long term even for the affluent EU, much less for a poor country like Latvia.

Rēķis's vision for sustainable rural development in Latvia was emphatically postagrarian:

Latvia currently has too much agricultural land. . . . Here the land quality is not high enough, the infrastructure is not good enough, nor the resources; the distance, transportation, all the costs are enormous. It is simply not cost-effective to produce more here, to produce for export. . . . I would guess that the total agricultural land could be reduced at least by half. Even if, of course, it goes against the rural—that is, the Latvian—lifestyle. Concentration of the rural population should be promoted. . . .

It's the story of the little old lady in the middle of the forest. Do we need 10,000 little old ladies struggling out there, and the mail carriers struggling, and the [state electric utility] struggling, and the bus drivers struggling to drive those horrible roads, so that she can live there another ten or twenty years? They should be encouraged to concentrate. And if someone wants to go back to live in the forest, then let him pay for it himself. So that the state doesn't get saddled with the burden of supporting them all, at all costs, in precisely that place, precisely in the middle of that forest, where nothing really grows, where there are no roads or electricity and God knows what else.[58]

Against the entrenched Latvian discourse that equates farming with closeness to nature, Rēķis insisted that "from an environmental perspective there is nothing at all good about people living [in the forests or countryside]; they fragment the landscape, build roads, and so on." In his view, the state should promote the movement of rural dwellers to population centers, although not in the Soviet style of "piling them into trucks and carting them off to the village." Living in dispersed farmsteads and working the land, in other words, was not equivalent to good nature stewardship, nor was it a fundamental right and duty of authentic Latvian citizens. Rather, it was "actually a privilege to practice the 'Latvian' lifestyle in this way." In short, Rēķis and Rotbergs advocated a radical redefinition of Latvians' place in nature, from the traditional labor of agrarian cultivation to the postproductivist work of managing it for natural processes. Instead of subsidizing agriculture, whether conventional or low-impact, WWF hoped to spur Latvians to capitalize on the international demand for biodiversity, through ecotourism, green certification for the timber industry, and other approaches to diversifying the rural economy.

The narratives of biodiversity and sustainability are rooted in the scientific claim that neither natural processes (such as migration of species) nor the impacts of environmental degradation (such as pollution or climate change) respect national borders. The geopolitical corollary of this claim is that nature can be stewarded properly only through international cooperation. "No nation can achieve this on its own," declares Agenda 21, the manifesto of the 1992 Rio Earth Summit that ushered in the era of international environmental action, "but together we can—in a global partnership for sustainable development."[59] The global approach to protecting nature has been codified in a vast array of conventions, treaties, and other documents, most of which Latvia has duly signed. "Presently Latvian territory supports self-supporting populations of many species, habitats and ecosystems that have become endangered or extinct in Europe," states the National Report on Biological Diversity. "This fact alone increases Latvia's international responsibility to biodiversity preservation for all the European continent."[60]

In the post-Rio world, nature is defined by the border-crossing systems and processes that comprise biodiversity. Its value is determined by international norms, it is protected by international institutions, and it is enjoyed by international consumers. By the late 1990s, some Latvians were beginning to embrace this view of nature. If international treaty obligations were the stick, then international environmental assistance was the carrot impelling Latvians to globalize their ethnoscape.

Since Latvia's National Awakening, generations of agrarian nationalists, from Atis Kronvalds to the post-Soviet ethno-naturalist contributors to the *Almanac of Nature and History*, have endeavored to liberate Latvians from cultural imperialism by nationalizing nature. Today's advocates of biodiversity, extending Krišjānis Valdemārs's internationalist vision of homeland to the countryside itself, have sought liberation from the Soviet legacy of marginality and underdevelopment by globalizing nature. In one sense, biodiversity as Latvia's new postproductivist capital signals the unraveling of the ethnoscape. The undrained wetland, it could be said, is the very opposite of the well-tended farmstead. But at the same time, high nature-value farming entails precisely the intimate, labor-intensive relationship of the farmer and his homestead. Thus the ambivalent message of the postproductivist transition provided internationally authoritative ammunition for agrarian-minded conservationists in their defense of the traditional ethnoscape and allowed them to sustain the hope of embracing the global without rejecting the national. "We need not sacrifice our cultural-historical values in our mystical efforts to be 'like everyone in Europe,'" suggested a newspaper editorial. "Perhaps people will come here expressly to enjoy the peace and blessedness of our rural homesteads?"[61] Ecotourism can mean many things: from tramping through insect-infested marshes in thigh-high waders to relaxing in a quaint farmhouse over a cup of fresh milk. Agrarian conservationists like Priednieks and Kaminska envisioned ecotourism as one tool—along with environmental aid schemes tailored for agriculture—for maintaining Latvian nature as a peopled, cultivated landscape.

But the long-term prospects of this scenario appeared limited at best. On the one hand, while tourism is booming worldwide, ecotourism in Latvia was still a tiny niche market. Country Traveler, a Latvian association of providers of rural tourism founded in 1993 with twenty participating members, had grown to include one hundred farms by 1998, receiving over 6,000 visitors annually, 20 percent of them from abroad.[62] This represented significant growth for a nontraditional form of enterprise, but the economic impact had thus far been minimal. Given Latvia's northern latitude and short summer season, ecotourism is unlikely ever to support many people. As a foreign lecturer bluntly observed, Latvian policymakers "hugely overestimate" tourism's ability to spread wealth beyond Riga and its immediate environs: "It's expensive to get here, the roads are slow and very little English or German is spoken outside Riga."[63] Meanwhile, it was by no means clear whether European conservationists would ultimately win their battle to increase support for high nature-value farming. The deepening of liberalization could easily bring, instead, a decline in aid to both farming and conservation. Thus analysts of the postproductivist transition wondered: "Are further attempts to 'green' existing systems of agricultural support either feasible or desirable in the long term; or should conservationists be preparing for a future in which government support to agriculture no longer exists, at least in its present form?"[64] It was an open question whether the EU would make a financial commitment to

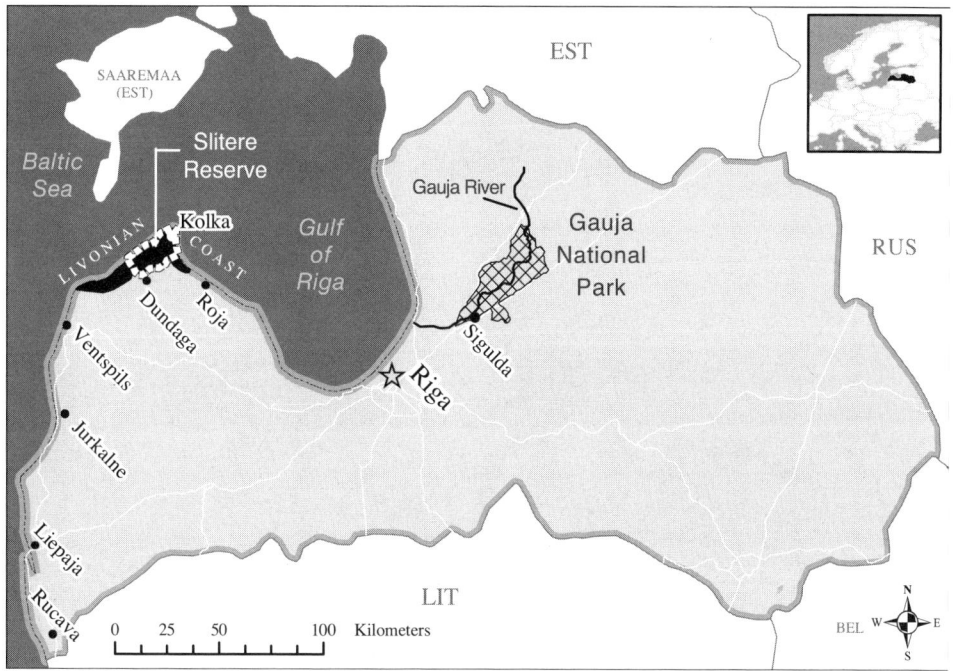

Figure 6. Map of Latvia with case study sites. Map: Robert D. Lopez.

maintaining the farmer as the "guardian of the landscape," and it was even less likely that Latvia, lacking a wealthy urban population, would be able to make up the difference.

As Latvia negotiated its encounter with the global narratives of sustainability and biodiversity, the tension between the inward- and outward-looking orientations was blunted by the ambiguity of these narratives themselves. Agrarian-conservationists hoped to have their global biodiversity and keep their ethnoscape too, but could this dream be realized for more than a small segment of the rural population? Would liberalization and integration into the EU ultimately force a retreat from the agrarian ethnoscape, as depopulation and land abandonment increasingly superimposed the wetland over the farmstead, the *viensēta*? We now turn to three case studies of Latvian participation in—and opposition to—Western assistance projects to examine in more detail how Latvian responses to the challenges of post-Soviet nature management and rural development were informed by competition between agrarian and internationalist discourses of place and identity.

Chapter 5

"Masters in Our Native Place"
The Politics of National Parks

"National parks would seem to be as much about the nature of national identity as about physical nature," observes Kenneth R. Olwig.[1] The world's first national parks were created in the United States as a manifestation of American cultural anxiety vis-à-vis Europe.[2] Since the establishment of Yosemite and Yellowstone in the late nineteenth century, national parks around the world have been particularly revealing sites for observing the deep linkages between identity discourses and constructions of nature. Thanks to their weighty symbolism and public prominence, national parks bring to the fore more acutely than other protected nature areas the tension between notions of nature as wilderness and as lived-in landscape, and between global and local environmental narratives.

In the seminal American cases, national parks were imagined as "empty" spaces devoid of humanity's footprints, and productive uses were strictly limited or excluded from park territories. But in most cases, writes Mark David Spence, "uninhabited wilderness had to be created before it could be preserved."[3] Preserving wilderness often required erasing these territories' actual history of centuries of habitation and exploitation by Native Americans and, all too often, their violent removal from these ostensibly "virgin" lands.[4]

Like so many other American creations, the national park idea took root around the world, and along with it the American understanding of parks as empty wilderness. Perhaps the best-known examples of this dissemination are Africa's first national parks, established by European colonists who imagined Africa as an untouched Eden and superimposed their own romantic wilderness ethic onto indigenous landscapes of labor and culture. In South Africa in the 1920s, the campaign to create Kruger National Park united British and Afrikaaner whites, across class and ethnic divides, around a common heritage of love of nature while excluding local subsistence hunters.[5] The creators of Serengeti National Park in British Tanganyika in the 1940s had to repress the fact that the Maasai and other native Africans had long been dwelling in and exploiting this ostensibly primeval wilderness; local residents were either forced out of park territory or required to live and farm only according to "traditional" practices, as defined by British authorities.[6] In the 1960s, the establishment of Matopos National Park in Zimbabwe engendered a clash between white and black understandings of heritage and of the meaning and value of the protected landscapes. As Terence Ranger reports, the whites' ecological notion of "a uniquely fragile eco-system which needed to be protected against man and his domestic animals" was pitted against indigenous communities' sense of a landscape "created over centuries by human use," in which their own presence "was sanctioned by history, religion and custom." Whites successfully deployed ecological narratives to ban native farmers from park territory, but these were reinforced by romantic appeals to preserving "the old Africa"—the Africa of vast Edenic expanses in which the white hunter could find himself alone in the wilderness.[7] Today, the era of colonialism may have passed, but indigenous communities and their advocates often critique Western-sponsored conservation projects in the developing world as neocolonial strategies for territorial control, similar in effects if not in means of implementation to the earlier efforts of colonial authorities. Nowadays, they note, the romantic narrative of Eden is often replaced by the scientific narrative of biodiversity, but both have been used to trump local values by Western elites seeking to control nature on a global scale.

In colonial Africa as in nineteenth-century America, parks were shaped by discourses of national identity: by myths of the rugged outdoorsman, memories of heroic military campaigns, and others. But they were also born of the desire of fledgling states for status and recognition on the international stage. Indeed, since the Wilsonian era of national self-determination, possessing a national park has become a symbolic marker of full statehood. As Jane Carruthers notes: "Whereas no stigma attached to a country before 1914 if it did not have any national parks, after 1919 it was shameful not to have established such institutions."[8] Ranger links the ultimate success of the white ecological vision at Matopos to the claim that allowing native residents to impinge on the park's status as a biodiversity preserve would bring international discredit to Zimbabwe. Thus while national parks can be seen as the ultimate institutionalization of the ethnoscape, shaped by potent myths about the nation's character, history, and relationship to the land, they also reflect pressures to conform to international norms and values.

Two international assistance projects in the late 1990s brought Latvians face to face with the question of whether the management of national parks should give precedence to local meanings and traditional uses of the land or to global norms and international status. At Gauja, Latvia's flagship national park created in the 1970s to protect the Latvian agrarian ethnoscape, internationalist-minded conservation officials and their Western advisers attempted to redefine the park in terms strictly focused on biodiversity, but they met with stiff resistance from their agrarian-minded peers. And in northern Kurzeme, internationalists seeking to create a new national park were stymied by local residents embracing a staunchly productivist agrarian understanding of nature. In neither case were Westerners imposing particular park boundaries or management regimes upon locals, but rather Latvians themselves were advancing competing visions for the parks in response to Western ideas and assistance. Where internationalists saw a savvy embrace of the globalization of nature, agrarians saw self-imposed neocolonialism.

Gauja National Park

Preparations for the 1973 centennial of the founding of Yellowstone, the world's first national park, sparked lively debate in nature management circles worldwide. In Latvia, too, international ferment inspired conservation professionals to begin discussions with their Soviet colleagues, especially in neighboring Estonia and Lithuania, about establishing national parks in their republics. The Soviet Union boasted a great many protected nature areas of various kinds, but none was designated a national park because of the politically objectionable character of the label. According to the geographer Aija Melluma, a key participant in these discussions, Soviet authorities initially rejected the label *national* (Russian, *natsional'nyi*; Latvian, *nacionālais*) in favor of the more orthodox *people's park* (*narodnyi*, *tautas*). They finally conceded after the idea's promoters fortuitously located a passage in Lenin's works calling for the eventual establishment of "American-style national parks" in the USSR.[9] Thus the Soviet Union's first national parks were born: Estonia's Lahemaa in 1971, Latvia's Gauja and Georgia's Tbilisi in 1973, and Lithuania's Augštaitija in 1974.

The ethno-nationalist significance of these pioneering Soviet national parks was never overtly acknowledged under Soviet rule, but the fact that they were established in precisely those republics that later led the ethno-nationalist drives for independence suggests that the word *national* was indeed more than a mere label. Certainly in the Latvian case the founding of Gauja was, like the great tree liberation movement, an example of Brezhnev-era crypto-nationalist dissent under the guise of landscape protection. Latvia's flagship national park was designed explicitly to protect not only natural habitats of great scenic value and biotic diversity, but also the agrarian cultural landscape, or Heimat, of fields, meadows, and farmsteads. And according to Melluma, its principal designer, the park's establishment precipitated a "great sense of national uplift" and a tremendous outpouring of social interest and support.[10]

The centerpiece of the park's 92,000 hectares, located in the central Vidzeme region fifty kilometers northeast of Riga, was the valleys of the Gauja River and its tributaries, whose steep sandstone bluffs and forested ravines had sheltered from aggressive human encroachment unique geological formations and a wealth of plant and animal life. Along this stretch of the Gauja, as Melluma noted in 1974, "nature has still to a large degree remained primordial and has been little transformed by human economic activity."[11] But the park designers also drew the boundaries so as to include wide swaths of agrarian landscape and many cultural and historical sites, such as castles, archeological remains, and caves celebrated in folk legend. For Guntis Eniņš, Heimat enthusiast and great tree liberator, the park's "real magic and value" were associated simultaneously with "living, untouched nature in the Gauja's inaccessible ravines" and with "history's distant and yet nearby breath in the ancient buildings."[12] For Melluma, the park territory was interesting not only for its geological and biological features, but also as a place to study "the 'footprints' of human economic activity in the present-day landscape."[13] Indeed, Gauja was one front in Melluma's broader campaign during the 1970s to steer nature protection in Latvia away from the exclusionary orientation of Soviet practice back toward its roots in German-style protection of the Heimat.

From Strict Nature Preserve to Protecting the Heimat

As discussed earlier, early conservationist ideas in the Russian Empire were informed by German activists' focus on the cultural landscape. In the Baltic provinces, the Riga Society of Naturalists was founded in 1845 by Baltic German amateurs, and one of the empire's first nature reserves was established in 1912 on Moritzholm Island (in Latvian, Moricsala) in northern Livland.[14] In 1923, the government of the newly independent Latvian republic enacted a law to protect monuments of nature (a category borrowed from the Germans), spurring the establishment of over twenty reserves of various types during the interwar period. "Prior to their forced annexation to the USSR in 1940," notes Philip R. Pryde, "the Baltic republics had achieved an admirable conservation record."[15]

The Heimat approach to conservation remained prominent in Latvia throughout the 1920s and 1930s. Not only naturalists but also agronomists and foresters called on their countrymen to "cultivate beauty in Latvia's forests and fields." "If we want to be real foresters," declared a contributor to the foresters' association

Figure 7. Gauja National Park tourist postcard. Photo: Indra Čekstere.

journal, "then we must look at the forest not from the practical side alone, but must search for beauty and relaxation in it as well."[16] In the geographical encyclopedia *The Land, Nature and People of Latvia* (1936), monuments of nature were defined as including not only geological formations, rare species of fauna and flora, reserves, and parks, but also objects of

> ethnographic, ritual, historical or cultural-historical interest, as for example sacrificial trees and rocks; . . . idol and altar hills; castle mounds; trees, lanes, and groves planted by notable national personages or planted to commemorate important events, battlefields, and the like. . . . These natural objects often no longer are in any way primeval or untouched by humans, which is the first requirement of a monument of nature in the true sense of the word; often they are fully man-made, as with tree-lined drives, parks, memorial groves.[17]

In Russia, by contrast, conservation practices diverged from the German *völkisch* pattern in the 1910s and 1920s, when a group of pioneering ecologists began promoting the establishment of strict nature reserves (*zapovedniki*) that would be protected from all human encroachments and would serve as models of purportedly "pristine" nature for scientific research. Their goal of strict protection was brutally undermined from the 1930s on by the Stalin regime's radically technocratic efforts to transform nature in pursuit of material progress.[18] Nonetheless,

Soviet conservation practices were deeply influenced by the notion of protected nature areas as places from which humans should be excluded.

In Latvia, too, after the Soviet annexation, the cultural landscape approach gave way to efforts to establish relatively "inviolable" nature reserves. The creation of protected nature areas accelerated, and by the late 1970s Soviet Latvia boasted 150 general and specialized reserves, operating under strict protection regimes and devoted to scientific research.[19] At the same time, however, ecologists had begun to argue that prehuman, "pristine" nature no longer existed in the Soviet Union and that current biota were, in fact, the product of centuries of hay-mowing, pasturing, and other human activities. "Semi-natural, human-transformed systems" should therefore be granted the same protection that ostensibly "pristine" reserves had hitherto enjoyed.[20]

In Latvia the geographer Melluma, arguing that not only relatively untouched nature reserves, but also "areas where nature maintains the footprints of various human activities" were of scientific significance, called for a new "protected landscape area" category in Latvian conservation policy.[21] While permitting agriculture and other uses of natural resources, this designation was intended to cover "old country parks and arboretums, protected millponds, . . . old field systems, . . . winding roads, and various other features created by human hands, which vividly demonstrate the history of the relationship between humans and nature."[22] In other words, Melluma sought to preserve the elements of the agrarian ethnoscape that, in her view, represented "the essential prototype of today's Latvian landscapes."[23] At the new Gauja National Park, Melluma and her fellow planners defined the nature protection mandate very broadly to include not only natural systems and genetic diversity—as in a strict reserve or *zapovednik*—but also "the material witnesses to the centuries-old interaction between man and nature."[24]

Homeland studies were an important focus of park activities from the outset. Indra Čekstere was hired as the park's ethnographer in 1974 and spent the next decade carrying out a detailed ethnographic inventory, visiting most of the park's 2,000 historic farmsteads (*viensētas*) and gathering "descriptions, diagrams, and photographs of the most valuable *viensētas*, as well as of outstanding monuments of vernacular architecture."[25] Čekstere supplemented her data with materials from local museums of regional history to compile a database of cultural and historical monuments in Gauja National Park. Residents of inventoried buildings were informed of the value of their buildings and requirements for their maintenance. Čekstere sought to "fix" the history not only of vernacular architecture but also of regional craftsmanship, recording oral histories and filming traditional craftsmen at work.[26] In 1982, her ethnographic informants came together for an "unexpectedly well-received" celebration that featured a performance by the park's own folklore ensemble, formed the previous year with the mission of "propagandizing the riches of Latvian culture, its ethical and aesthetic values, and understanding of nature" through the folk songs of the Gauja region.[27] At the ancient village of Āraiši, a particularly rich historical and archeological site, park authorities launched a project in 1983 to preserve the area as an example of "the characteristic Vidzeme landscape with its groves, old roadways, and lakes," including restoration of all existing farmsteads. Park staff joined local residents and members of the Latvian SSR Society for Protection of Natural and Historic Monuments in making regular inspections of the Āraiši territory. Efforts were made to preserve valuable objects in each of the park's other townships as well, including both old farmsteads and the newer ones created through the "bourgeois agrarian reform" of the 1920s.

Gauja staff members understood their mission as including efforts to rekindle Latvians' fading awareness of their local history and folk traditions, and to protect the embodiments of that history and "mentality" in the cultural landscape against the onslaught of Soviet rural policies. Throughout the 1980s, park officials invoked Čekstere's ethnographic data to "halt the demolition or thoughtless remodeling" of many historic buildings, and they used zoning and management regulations to curb objectionable land use practices such as intensive land reclamation, unplanned construction of summer cottages and tourist facilities, and the mass expansion of private farm plots during the campaign to boost food supplies in the late 1980s.[28] The park's first director, Gundars Skriba, recalled on the occasion of Gauja's twentieth anniversary in 1993: "Like a magnet, the Gauja Valley with its naturalness, beauty, and many meadows and glades attracted . . . all manner of prominent and moneyed persons wishing to establish their garden plot, build their cottage, house, or sauna precisely here and nowhere else."[29] Park managers successfully rebuffed many of these attempts, making Gauja National Park another bulwark against Sovietization of the Latvian Heimat.

Shedding the Ethnoscape?

After independence, the status of Gauja as a large, complex protected area encompassing both "primordial" nature and agrarian ethnoscape came into question. Because Gauja was established during the Soviet period without regard to prior property rights, post-Soviet restitution created thousands of new private landowners within the park's boundaries, many of whom now struggled to eke out a living by farming and logging, or hoped to capitalize on the area's natural amenities by developing tourist facilities. Because of the original planners' desire to simplify external boundaries, the park also encompassed several small towns and densely inhabited areas. The zoning scheme thus comprised seven distinct zones, each with its own management regime: the strict nature reserve, natural landscape, cultural landscape, recreational landscape, light development and heavy development, and neutral or buffer zones. Management of such a complex protected area had become significantly more complicated in the post-Soviet era of private property rights and hard budget constraints. In 1998 the Latvian government initiated a project, with funding and technical support from Denmark, to draft a new management plan and regulations for Gauja.[30] The Danish project, as it was commonly known, became an important testing ground for modern environmental management techniques borrowed from the West.

During the Soviet period, studies of plant and animal life at Gauja were sporadic, unsystematic, and poorly disseminated. Maps were considered military secrets and off-limits to civilians.[31] In stark contrast, the Danish project employed

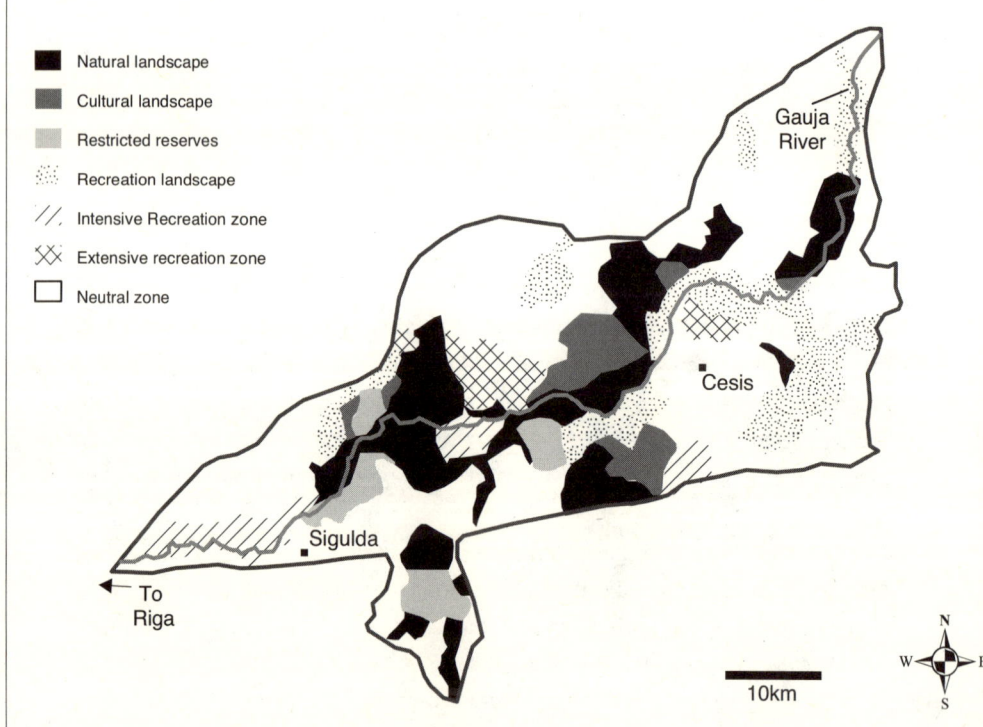

Figure 8. Map of Gauja National Park zoning. Map: Robert D. Lopez, adapted from Latvian State Forest Service and Holsteinborg Consult.

sophisticated Geographical Information Systems (GIS) technology to layer biological, geological, socioeconomic, and ethnographic data onto aerial photographs to produce multidimensional digitized maps for use in management planning. Project managers gave top priority to systematic ecological field surveys,
beginning with an inventory of "woodland key habitats," to identify areas of
greatest biodiversity value. These rationalizing techniques reinforced a notion of
nature as first and foremost a reservoir of biodiversity. The project team leaders,
Jānis Strautnieks, director of Gauja, and Aase Ostergaard, a Danish consultant,
advocated excising the heavily inhabited zones and most of the protected landscape areas to reduce the park's territory to its relatively untouched "core zone,"
comprising the river valleys with their forested banks and wet meadows. This reduction, they argued, would not only simplify park management and prevent
many potential conflicts with private landowners, but also move Gauja toward
compliance with international standards. According to the widely accepted IUCN
guidelines for protected area management, "At least three-quarters and preferably
more of [a protected area] must be managed for its primary purpose." For national parks (category II), this purpose is identified as ecosystem protection and
recreation.[32] With its acres of farmland, villages, and zones of light and heavy development, Gauja fell considerably short of this three-quarters standard. The
strict reserve and natural landscape zones made up only around 5 percent and 16
percent, respectively, of the park's total territory, while the buffer zone accounted
for 51 percent.

For director Strautnieks, the park's sovereign mission was to protect the diversity of species and habitats. "I won't talk to you about the fortress ruins," he told
me, dismissing the park's cultural and historical heritage as a "secondary matter."[33] Strautnieks hoped to achieve compliance with IUCN category II criteria in
ten years' time, if not by shedding the park's developed zones, then at least by
banning all economic activities other than recreation within the core zone. The
primary motivation for this agenda was, as he put it, "to keep up with the times, to
anticipate where Europe is heading." The influence of IUCN and its global norms
was "inevitably" growing throughout Europe, Strautnieks pointed out, and now
in Latvia, too, nature management was "being raised to an entirely new level, essentially based on Europe's demands—not due to local pressure." Now Latvians
needed to demonstrate their commitment to

> establishing a protected nature area system, or at least one protected nature area, in a
> manner that is understandable in the Western world. Understandable at least to the
> Scandinavians, say, who are right here nearby and who see what we are doing. Nowa-
> days there is much more contact, they travel here and wonder, and draw their conclu-

sions and fill up whole newspapers. . . . If we don't demonstrate this, then I think that we will not be able to hope for support from foreign governments and various foundations.[34]

For Strautnieks, having an internationally recognized national park was one "path toward [Latvia's] achieving integration" into Europe.

Compliance with global norms and values promised to enhance access not only to funding from international donors, but also to international tourism. "National parks make valuable 'green visiting cards,'" Stuart Franklin notes, and according to Strautnieks Gauja National Park was "the second most popular destination, after Riga, for international tourists. Gauja represents Latvia and its nature."[35] It was crucial, then, to provide the sort of nature that Western tourists wanted to enjoy. The key question for Valdis Pilāts, Gauja's senior staff ecologist, was this: "How does the Western tourist differ from the (ex-) Soviet tourist? What does each wish to see at Gauja National Park, and what deficiencies does each perceive?" Latvians, Pilāts observed, wanted scenic beauty and sites of folkloric significance and were distressed by a disorderly landscape, wondering: "Why are the parks' woods so cluttered, why haven't the dead and fallen trees been removed?" In contrast, "the Swede, the German, the Dutchman, the Dane—nearly all of them ask: 'Why is your forest so new and why does it look so cultivated? . . . And where is your old-growth forest?'" The problem, according to Pilāts, was that "in Latvia, it seems, primordial nature has never been celebrated. Only that which accords with human aesthetic canons, which is created or transformed by human hands, is valued." But the future, for Pilāts as for Strautnieks, lies with the Western understanding of nature:

> Some will say that Gauja National Park must serve our own people first and foremost, and that we need not imitate Westerners. That's fine, but let's look ahead. What will Latvia be like after ten, twenty years? How much of that insignificant "untouched" nature will we have left? What will the Latvian visitor to Gauja National Park wish to see after ten or twenty years?[36]

For Strautnieks and Pilāts, biology trumped ethnography. Gauja National Park should be internationalized by eliminating the elements that embodied the traditional Latvian relationship to nature as that of the good steward (saimnieks) to his farmstead (viensēta).

However, the reduction proposal met with such firm resistance from other members of the project steering committee—notably the park's "author" Aija Melluma and her colleagues from the environment ministry—that it was ultimately shelved. According to the Danish consultant Ostergaard, practical questions, such as whether sustainable land use planning could be adequately enforced outside the park's borders, were only minimally discussed at project meetings. Instead, Ostergaard lamented, the debate centered on "principles, not reality," which she considered a "waste of time."[37] What was at stake, though, was a fundamental clash between internationalist and agrarian understandings of the meaning of national parks.

Not only internationalists but also defenders of the agrarian ethnoscape were able to appropriate Western narratives to support their position. While the former emphasized biodiversity, the latter drew upon the sustainable development or "people and parks" paradigm: this was the idea that both biodiversity and local values can be protected by integrating conservation goals into international development assistance and by devising ways for residents to profit from sound nature stewardship through ecotourism, heritage tourism, controlled hunting, sustainable resource harvesting, and the like. "All this time I've been following the literature on national parks fairly closely," Melluma told me, "and basically even at those [IUCN] congresses, everyone talks about the national park as a development area."[38] She argued that Gauja should serve as a tool not only for biodiversity conservation in the core zone, but also for promoting sustainable development and land use planning in the region as a whole. The narrow focus on biodiversity cast humans in the purely negative role of defilers of nature, she maintained. This dynamic perpetuated the legacy of hostility between park authorities and local residents inherited from the Soviet-era exclusionary approach to management of protected areas. In her view, "The entire nature protection system should be humanized," and private landowners within park boundaries should be assisted in nature-friendly development planning, rather than hemmed in with countless land use restrictions. She lamented the elevation of biological values as the centerpiece of Gauja's management vision: "Here in Latvia the human influence is very ancient, and today it is very hard to determine what is purely natural and what has been shaped by humans."[39]

While Melluma used the current buzzwords of the Western paradigm—sustainability, stakeholder participation, natural dynamism—in making her case, her vision was grounded in the familiar Latvian construction of nature as a site of human labor and reservoir of history and national identity. A landscape is like an archive, she told me. "It can simultaneously contain very ancient elements—fortress ruins, say, which date from something like a thousand years before our era, along with the very newest elements." The "most compelling" aspect of nature protection for her was exploring this "archive" to uncover the historical evidence of the human impact on nature. While Melluma did not argue in explicitly nationalist terms, the natural heritage that she wanted Gauja National Park to embody and protect was not primarily the global nature of species and habitats but the agrarian ethnoscape.

Northern Kurzeme National Park

Latvia's second-oldest protected nature area was the Slītere Reserve, established in 1921 in the northern Kurzeme region, where the Baltic Sea meets the Gulf of Riga at the Horn of Kolka.[40] Slītere is an area of great natural diversity, situated in a remote and economically underdeveloped region. A wide range of forest and bog types support many Red List endangered and threatened species, and thousands of migrating birds—buzzards, eagles, ducks, geese, gulls, and passerines—fly over the Horn of Kolka each spring. Slītere's diverse topography vividly illustrates the geological history of the era following the Ice Age. Thanks to the area's remoteness, much of the greater Slītere ecosystem has been relatively untouched by human activities. Moreover, the Soviet regime had cordoned off the entire Baltic Sea coastal region as a militarized border zone. Nonresidents were denied entry, and travel between towns and villages within the zone was restricted and unpleasant even for residents. Nearly a decade after independence, roads in northern Kurzeme remained few and poorly paved, and a dearth of public transportation left many coastal residents largely isolated. Population growth was negative, and population density ranked among the lowest in the country—18 and 6 persons per square kilometer, respectively, for the two northern Kurzeme counties, Talsi and Ventspils, and as low 1.9 for Ance, Latvia's most sparsely populated municipality.[41]

Over the years the Slītere Reserve had been enlarged several times, growing by the late 1980s to 15,000 hectares and encompassing not only undeveloped forests and bogs, but also several coastal fishing villages. Unlike many other coastal areas, the villages of northern Kurzeme had not been emptied during the Soviet period. While individual fishing was banned, many residents found work at local fishing and processing collectives, and some of these enterprises stayed in business after independence and privatization. These villages are also home to one of the last remaining communities of Livonians or Livs, the Finno-Ugric seafaring people who were the first inhabitants of present-day Latvia but who were long ago assimilated into the Latvian mainstream. In 1991 this coastal territory was designated as the Livonian Coast, a protected area dedicated to preserving the near-extinct Livonian language and cultural heritage.

In the mid-1990s, spurred by the Finnish-supported Kurzeme Ecotourism Development Project and other Western aid initiatives, local and national environmental authorities began to explore the possibilities for promoting nature tourism in the area. In July 1996 Latvia's minister of environment and regional development proposed merging the Livonian Coast, Slītere, and several smaller nature reserves to create a new Northern Kurzeme National Park. The park's mission would be to protect both natural and cultural heritage and to promote tourism and sustainable local development; its territory would encompass Kolka township in its entirety and portions of Ance, Dundaga, Roja, and Tārgale townships. However, whereas the founding of Gauja National Park in the early 1970s had sparked an outpouring of public support and a "sense of national uplift," in the late 1990s the northern Kurzeme proposal unleashed an avalanche of hostility from local officials and residents.

On one level, this hostility reflected the peculiar dynamics of local politics in the area. Whereas Gauja had been created as a national park, Slītere had always been a strict nature reserve, exclusively devoted to scientific research and protection of natural values against human despoliation. During the Soviet period, recalled Elmārs Pēterhofs, a biologist and the long-serving director of Slītere, "the chief objective was creating a serious enforcement system, so that we would be in control of the entire territory."[42] Slītere officials had no mandate for promoting recreation or education and outreach; therefore, according to Gunārs Laicāns, mayor of Dundaga, where the reserve's headquarters are located, "If you asked twenty people on the street in Dundaga what the Slītere Reserve management does, they would all shrug their shoulders and no one would be able to tell you anything, not one word."[43] Park rules restricted local residents in their traditional subsistence activities of gathering berries, mushrooms, and firewood in the forests, making relations with park officials tense.

Some restrictions had been relaxed since the end of Soviet rule, but badly botched public relations efforts by Pēterhofs and national environmental officials exacerbated this legacy of hostility and mutual suspicion. Their clumsy and perhaps not entirely good-faith efforts at Western-style "participatory planning" fostered perceptions of a project that had been hatched in secrecy "at the top" and imposed upon unwitting local residents. Subsequent efforts to involve residents failed to dispel this simmering resentment, which, some park promoters claimed, township officials actively exploited to consolidate their own popular support. Local authorities' fears echoed a broader fear of losing political control over their townships through an impending nationwide reform of municipal and county government administration. Initial suspicions were compounded by locals' mistrust of the managers of the Livonian Coast, who were widely reviled for alleged financial improprieties, poor community relations, and failure to improve the material well-being of Livonian residents. Finally, local authorities feared that they would not be compensated for any loss of property-tax revenues resulting from expanding the protected area.

The inhabitants of these isolated, hard-scrabble fishing villages have a long-standing reputation in Latvia as taciturn and uncongenial, and this stereotype was frequently invoked by both local and outside observers to explain why it was so difficult to have productive discussions of the park proposal. Indeed, a particular constellation of local legacies, rivalries, and resentments transformed the national park debate into a bitter "war of all against all," as one local observer described it. But at the same time, the battle lines reflected familiar discursive divisions. This was not only a struggle over authority, but also a struggle between two competing developmental visions and notions of Latvians' proper relationship to the land.

For Slītere director Pēterhofs, as for Gauja's Strautnieks, the chief value of the proposed national park was its natural diversity, and the area's greatest developmental potential derived from Westerners' interest in that natural diversity: from the recreational and ecological value of the region's undeveloped beaches, its distinctive "landscape of fens and dunes," and above all its biological variety. Observed Pēterhofs:

> During the spring migratory period, there are days when 150,000 birds fly over the Horn of Kolka. Since Latvia's border has become more open, not only specialists but also so-called nature lovers come to observe them. In the Nordic countries it's like a movement—thousands of people travel on Saturdays and Sundays to bird-gathering spots, to observe them and identify species. . . . Both Latvian and foreign specialists agree that the Horn of Kolka is the most significant bird-gathering spot in northern Europe during the spring migration.[44]

Because of this Western interest, he maintained, "the Horn of Kolka could be a good site for tourism and money making, if only the Kolka residents could understand the tremendous business opportunities available in the service sector." The proposed national park was a crucial element of Pēterhofs's Valdemārian developmental vision:

> I think the chief stimulus [for economic growth] could be the opening of passenger terminals at Ventspils and Roja. . . . From [the old lighthouse in the village of] Slītere we can see Saaremaa island [in Estonia]. How everything is developed there! Truly at the highest level—ecotourism, all this tourism from Scandinavia, domestic Estonian tourism. And here it all ends. That's it—it goes no further. If the ferry traffic were extended to Roja, for example, or directly from Denmark and Sweden to Ventspils, then the flow of money would be opened in this direction.[45]

In Pēterhofs's view, indeed, ecotourism offered "the only development possibility" for this economically "frozen" region, with its aging and shrinking population.[46] The productivist developmental alternative was a hopeless dead end. In the spring of 1997, at a time when Kolka had no unemployment and was drawing over five hundred workers from other townships to its fish processing enterprises, Pēterhofs declared presciently:

> The welfare of Kolka, which is based on fisheries, is as insubstantial as a see-through nightgown and can vanish in an instant, with shifts in the market demand for processed fish in Russia, the sole export destination. Furthermore, people are cruelly overworked in fish plants for a meager 40–60 lat salary, and so you can't even talk about prosperity, but only about eking out a basic existence. The tourism business, in contrast, is an incomparably more civilized and promising income source.[47]

In the fall of 1998 the Russian economic crisis devastated Latvia's coastal fisheries, sending unemployment levels to 25 percent in Kolka by January 1999.

Like Gauja's Strautnieks, Pēterhofs stressed the importance of conforming to international standards for nature protection. "If we count up all our protected nature areas, we look very bad, very impoverished in the European context," he lamented, but a significant new protected area would enhance Latvia's international prestige.[48] Moreover, he told me, "Anyone who comes from Europe has a notion of what a nature reserve is and what it should be like," and Slītere Reserve, with its villages and zones of intensive recreation, did not meet these expectations. The core of the new national park should be its areas of "primordial" nature, he believed, particularly its natural and seminatural forests. Of particular interest to Western observers was an area burned in 1992 in one of Latvia's largest forest fires, because the Slītere management, flying against the country's tradition of landscape cultivation, was enforcing natural regeneration instead of replanting. Where local residents saw only an ugly wasteland and a shameful squandering of firewood, visiting Swedish scientists saw a valuable site for scientific study, declaring enthusiastically: "We believe that the Slītere forest fire actually increased the protected area's value."[49] As Pēterhofs and other park promoters frequently pointed out, a national park was an internationally recognized designation and, as such, would open the door to international assistance and funding.

"Where Do the People Go?"

At the same time, though, the new park was to encompass not only Slītere Reserve but also the Livonian Coast protected area, with its mission of cataloguing and protecting vernacular architecture, folklore, and other cultural elements of the Livonian ethnoscape. The original Northern Kurzeme National Park concept was drafted by Aija Melluma, then serving as a Livonian Coast board member and advisor to the environment minister. Melluma conceived of the park as protecting cultural landscapes as well as "a unique and relatively untouched ecosystem, coastal landscape, and natural diversity." An early draft of the management plan called for negotiating with local landowners and finding "ways of subsidizing the traditional land use forms," in order to preserve "the most valuable traditional rural landscapes together with the cultural heritage architectural structures."[50] As at Gauja, Melluma's management vision was one of sustainable development rather than exclusionary nature protection, and Pēterhofs and other park boosters enthusiastically argued that national park status would do more to encourage development than the park's current status as a nature reserve.

However, many residents and local officials—particularly the mayors of Kolka, Ance, and Tārgale townships—were deeply skeptical about the national park's developmental potential. The mayor of Kolka, Maija Rēriha, by many accounts the ringleader of the opposition, insisted that given the current poor quality of drinking water, roads, and other infrastructure in the region, "visions of flourishing tourism are fruitless dreams." She predicted that "the anticipated international funding would go to the researchers of flowers and Livonian folk songs, and the possible income from tourism—to the national park."[51] Rēriha and her allies envisioned their region's future flourishing only by strengthening fisheries, forestry, and farming, poor soil conditions notwithstanding. "Everything possible must be done to develop the traditional occupations," she insisted. "If the operations of the local fish processing firms should be limited, that would be a tragedy for Kolka dwellers."[52] According to Tārgale's Mayor Biruta Ēce, agricultural price supports "ought to play the chief role" in her township's development. Like other agrarian protectionists, she opposed rapid accession to the EU:

> We want to imitate Europe in many ways, but I don't know whether we can always pull it off. Really, as far as that European Union accession—well, at the top they make decisions and talk about all sorts of things, but it doesn't trickle down to us. . . . We're not ready to join the European Union, we're not developed enough to go with their prices. . . . Country people simply think we're not ready yet.[53]

In sharp contrast to Pēterhofs, with his vision of the northern Kurzeme coast as a hub of international transit, park opponents were suspicious of outsiders and foreign influences. At the first, ill-fated public meeting in 1997, a journalist reported the allegations of many local residents "that this performance/meeting was orchestrated by a person or persons with an interest in the anticipated foreign investment capital; others maintained that the national park project would allow would-be builders of manors and castles into this heretofore off-limits territory."[54] Many residents scornfully dismissed the idea of grounding local development in the Western appreciation of nature as wilderness. "What do we have to show others?" asked a public meeting participant. "It's a pigsty here, a jungle, everything is overgrown."[55]

But unlike many rural dwellers elsewhere in Latvia (including many in the town of Rucava, considered in the next chapter), the northern Kurzeme folk also rejected the prospect of marketing their own ethnographic heritage. "What will happen if my goat leaves droppings on the trail where the gentlemen have to walk to enjoy the view? And I myself have to wear a folk costume and stand at the gate? Perhaps I will earn a few pennies?" asked one audience member. "The Livonians are now likened to Indians who live on reservations and are brought cheap liquor and prostitutes," declared another. "They don't have to work or fish, but only watch with clouded eyes."[56] Rēriha concurred: "The Livonian and the Latvian are proud, they don't have the mentality to sell themselves for money and display something. Anything else, but that—never!" On the other hand, "That we know how to produce and that we make a very good product, that much is clear."[57] "Rationally we understand and support [the park idea]," explained a retired lumberjack. "But if only the rich tourist will feel free here, while we ourselves won't be able to set foot in the forest or the sea on account of the many restrictions, then that's no kind of life. We have to go to the forest for firewood, to the bog for berries, to the sea for fish. And we have to cut hay for our livestock."[58]

The primary concern of those who opposed the northern Kurzeme park, voiced repeatedly despite assurances to the contrary, was that the land use restrictions in effect in Slītere Reserve would be maintained and even tightened in the national park, and area residents would be forbidden to engage in their traditional subsistence activities of gathering firewood, grazing livestock, and picking mushrooms and berries for sale at markets and roadside stands. In defending these traditional land use rights, park opponents invoked the familiar notion of the laboring Latvian *saimnieks* as an innately good steward of nature. "The ancient chronicles and documents indicate that the Kurzeme coast has been inhabited for over seven or eight centuries," observed Laimdota Sēle, a local reporter who vigorously opposed the park proposal, adding that these inhabitants "have always been wise neighbors to the unique landscapes, the diverse plant and animal world, and all have survived together up to our time."[59] First and foremost, however, opponents of the park defined their relationship to nature as that of producers and masters of the land, on however modest a scale. As one internationalist park promoter, Livonian Coast director Sīlis, lamented:

The locals view everything connected to trade as speculation, ever since Soviet times. Everything connected to services is just looking for Easy Street. Everything connected to higher education—that's a lost person, because he no longer works. But what does it mean to work? To work means to produce . . . to catch fish, produce agricultural goods. Maybe you can work in factories, too, but that's not the real thing, because the nation's base is of course in the countryside. Farmers and fishermen—those are the real thing, even if they will only be 5 percent in Latvia, they will still be the nation's foundation.[60]

Park opponents feared that local people, prevented from deriving direct material benefit from nature, would have no place in nature at all. Thus they perceived the national park idea in very stark terms as promoting nature protection over human interests. Kolka's Mayor Rēriha complained that the draft management plan "gives a great deal of thought to the plant world, to the forests, the bogs. . . . There is even a phrase that says . . . that these fishing villages are important for forests, plants, birds, insects, and all other kinds of biological diversity, but people are not mentioned at all. . . . Thus we see that there is no place anticipated here for people."[61] She predicted that over time, the stricter regulations of the core zone would be expanded to inhabited zones as well: as these areas "grow over with forests, birds, insects—people will be pushed out and a stricter zoning will be declared." Rēriha was unmoved by park boosters' assertions about sustainable development. "Where do the people go?" she asked. "What do they do, how do they live?"[62] As the journalist Sēle put it: "We have the Red List of flora and fauna species threatened with extinction; everything possible is done to preserve these natural wonders for the future. But where is the Red List in which the Livonian is listed? . . . The most favorable conditions are created for the survival of plants and animals, but these conditions are stolen from humans, the native inhabitants of this land."[63]

Fears of local residents being pushed off the land and threatened with extinction were rooted in a very real history of displacement and exile. As locals often pointed out, the first exile occurred in 1944 when coastal villagers were ordered by the occupying German forces to evacuate inland. Some residents fled to Sweden during the war; others were deported after the Soviet annexation. As with the rest of the Kurzeme coast, Soviet rule transformed the region into a vigilantly patrolled military outpost, as described by a journalist:

It is estimated that military personnel outnumbered civilians tenfold in this closely guarded corner of the socialist empire's western border. Radar facilities, tank and coastal artillery divisions of the border patrol were more numerous than villages. The broad, straight road running parallel to the sea was built in the sixties expressly to meet the army's tactical needs, and the locals called it the Khrushchev road.[64]

Movement in the area was controlled, residents had to carry passports even when traveling between neighboring villages, and access to beaches was prohibited outside the few designated public areas. The Slītere Reserve was widely perceived as yet another dimension of the restrictions and prohibitions imposed by an occupying regime, as well as a playground for the powerful elite—"more a place for the big men to enjoy themselves," according to Mayor Laicāns of Dundaga. "It was, as they say, a state within a state."[65]

Soviet rule had also brought nationalization of private property, and the post-Soviet restitution process was slow and problematic in northern Kurzeme. Because the interwar land survey had been cut short there by the outbreak of World War II, many parcels lacked clearly documented boundaries; but more important, in many cases restitution of seized land had been encumbered by the presence of Slītere Reserve. The Slītere administration was negotiating with landowners to exchange properties in the core zone for parcels in less ecologically significant areas, but these negotiations had been complicated by uncertainty over the future extent and borders of the park. Residents thus felt that, after having been exiled from their own land for half a century, they were now being threatened by a new exile in the form of Northern Kurzeme National Park. "What if I've been living all this time among eternal prohibitions," wondered Tārgale's Mayor Ēce, "and now finally it's my property, now my land is finally restituted, and then they put the park on top of it?"[66] Gundars Berkholds, head of the Kolka Livonian Society, expressed the same fear:

> Let's say I own some land, and some kind of trees are growing on my land, maybe an oak grove or something, and that will be a landscape. Will I be able to work there, and how will I be able to work there, even though the land is mine? Will the state buy it from me? . . . Where will we be allowed to drive, to walk, how will all that be regulated?[67]

After decades of being "very much hemmed in" by the military regime, he said, people wanted "to feel freer." Former landowners and their heirs "cannot return to the land of their ancestors," declared the journalist Sēle indignantly, adding: "What does this remind you of, readers? Perhaps the already partially forgotten times when Latvia's citizens were denied access to the sea along nearly the entire length of the Kurzeme coast?"[68] Residents feared that emparkment might not only decrease their access to the forest but even unleash a new round of expropriations.

From the perspective of the local opposition, the proposed Northern Kurzeme National Park hung "like a sword of Damocles" over the heads of coastal residents, threatening permanent underdevelopment and yet another foreign-imposed exile from their native lands.[69] Because of land use restrictions, as Mayor Rēriha put it, "the local person was no longer master in his own native place—*saimnieks savā dzimtajā pusē*."[70] Her poignant phrase evoked the deeply resonant separatist slogan of the glasnost era, which cast the struggle for independence from Soviet rule as Latvia's regaining its *saimnieks* and Latvians becoming once again masters in their own land. Her words also recalled the heroic *saimnieks* of Latvian literature and popular myth, who must carve a rugged livelihood from an unforgiving environment through his own productive labor. The hostility to the park thus combined the notion of liberation from outside dictates with the agrarian valorization of productive labor. Opponents desired the freedom to own and use land as they wished, unfettered by government restrictions, but they also felt entitled to government support in realizing their own unapologetically productivist vision of development.

Slītere director Pēterhofs firmly rejected the claim that establishing a park would "exile" local landowners by imposing stricter land use controls. Those living inside the park would continue to be allowed to gather materials and graze livestock; indeed, he insisted, "the more herbivorous livestock there are, the more meadows will be mowed and the more beautiful will be the natural landscape, which is one of the chief requirements of a national park." And he predicted, "As tourism develops, fishermen will be able to smoke their fish and sell them to tourists and vacationers in prescribed places."[71] At the same time, however, Pēterhofs, who had worked at Slītere Reserve throughout his entire twenty-year career, appeared not to have fully internalized the new Western notion of local residents as stakeholders rather than enemies. Thus he bemoaned the resurgence of individual fishing in the late 1980s:

> Fishermen felt that now they could drive across the dunes anywhere that was convenient for them. . . . The dunes are being destroyed, and I can guarantee that fishermen have destroyed more in these few years than the Soviet army during the entire postwar period. . . . People have destroyed so much that nature protection professionals now are trying to remove the most unique values from public use, in order to preserve them.[72]

Pēterhofs thus implicitly rejected the notion of the Latvian *saimnieks* as an instinctively wise steward of nature and of farming, fishing, or lumbering as closeness to nature.

While he paid lip service to the idea of national parks as zones for fostering development and interaction between human beings and nature, Pēterhofs's comments suggested that he remained wedded to the exclusionary paradigm. Melluma noted this dissonance, observing that as the draft management plan was revised over time, "it gradually began to lose all the emphases and aspects connected with development—it was once again transformed into a stricter protected area, which is inappropriate [for a national park]. If there's no discussion of the development of the fishing villages, then there's no point in talking about a national park at all! . . . You can't turn it into an empty territory, that is not our goal."[73] Despite her own intentions in designing the original management plan, Melluma feared that there might be some grounds for local residents' perceptions of the proposed park as inimical to local development.

Both the Gauja and the northern Kurzeme cases represented at least a temporary victory for defenders of the ethnoscape and a defeat for the internationalist, biocentric understanding of national parks. By late summer 1999, Strautnieks and his allies had indefinitely shelved the plan to reduce Gauja National Park to its biologically valuable core zone, and Pēterhofs was negotiating an alternative proposal to create a much smaller national park within the existing boundaries of Slītere Reserve. Slītere National Park was formally established at the end of 2000, and when I returned for a brief visit in June 2002, several new nature trails had been completed and a small museum was being set up in the newly remodeled Slītere lighthouse, but no outreach work had yet been undertaken to publicize the area's new status or increase tourism.

Internationalists at Gauja and northern Kurzeme embraced the biocentric view of nature and the transit-oriented vision of Latvia's developmental destiny, identifying biodiversity as a more valuable trading currency than farm goods, traditional farmsteads, or archeological ruins. But at the same time they often expressed a strong attachment to the "nation of farmers" concept. While Strautnieks feared that the failure of Latvia's flagship national park to meet Western standards could hamper Latvia's "return to Europe," he also believed the government should subsidize farmers to maintain agrarian landscapes, though not within the borders of a national park. So, too, did his westward-looking colleague Pilāts, who maintained that Latvia needed its farmers. "If you look at this nation, and where it gets its human resources from, then I want to say that it's primarily from the countryside. As a biologist—the valuable biological material comes primarily from the countryside."[74] Dundaga's Mayor Laicāns, while a strong supporter of the proposed Northern Kurzeme National Park, similarly affirmed that farming was the basis of Latvia's nationhood and its developmental destiny:

> I put the countryside in first place as the sustainer of our Latvianness in a positive sense, because if we study all of history and study the great figures that Latvians have had, then it turns out that all of them, at least the majority, have come from the countryside. . . . The countryside is what will also determine our future employment, because you can't simply live in the countryside and enthuse about something. . . . And that basically means agriculture, and that is how it has to be; and if we do not understand that, then we will very quickly head toward destruction.[75]

Strautnieks, Pilāts, and Laicāns essentially hoped to compartmentalize the ethnoscape, marketing some parts westward as biodiversity, while keeping a healthy portion in agricultural production as "the sustainer of our Latvianness."

The agrarian-oriented Melluma, too, embraced sustainable development as a mechanism for simultaneously preserving and globalizing the agrarian ethnoscape. She wanted to stress preserving the Heimat as an "archive" of Latvian national history, but at the same time she welcomed international ecotourism as a means of ensuring the viability of living cultural landscapes. In constructing nature primarily as a reservoir of cultural identity, Melluma reproduced the discourse of the great tree liberators and the folklore enthusiasts of the *Almanac of Nature and History*.

This ethno-naturalist narrative represents only one strand within Latvian agrarian discourse, however; the northern Kurzeme park opponents gave voice to another more purely productivst strand inherited from both interwar peasantist discourse and Soviet Marxism. Kolka's Mayor Rēriha and her followers were fighting not for the ethnographic Heimat of Kronvalds, *Indrāns*, and Ziedonis, but rather for the laboring landowner of Ulmanis, Jaunsudrabiņš, and Vilis Lācis. In their thoroughly inward-looking developmental vision, extraction of resources was the only dignified livelihood for local people, despite the region's poor soil and remote location and—even after the collapse of the Russian market—the risky nature of the fisheries business. Neither wilderness tourism nor ethnographic heritage tourism was an acceptable alternative. For them, creating a new park was not a matter of generating new value in a devalued space, but rather of imposing a foreign and undesirable value onto the local population "from the top." The national park was simply a mechanism for depopulation, for emptying the landscape even more effectively than the Soviet army had done. For the agrarian opposition in northern Kurzeme, conserving biodiversity represented nothing more than a new stragegy for territorial control, and biocentric internationalists were merely the tools of Westerners seeking to exploit Latvia as another Eden of biodiversity, albeit on a less grand scale than, say, Serengeti.

The northern Kurzeme case thus starkly exposed the difficulties of attempting to wed the notion of farming as closeness to nature to the westward-looking bio-centric paradigm. Under Soviet rule, the notions of being stewards of nature and masters in one's native place were woven together in crypto-nationalist defense of the ethnoscape. In post-Soviet Latvia, it remained to be seen whether agrarian na-tionalism could underpin a viable strategy for protecting nature in an era of Euro-pean integration. Could *national* support be mustered for nature defined in *inter-national* terms? Or was a national park created in the image of global biodiversity an oxymoron, even a form of self-imposed cultural imperialism?

Chapter 6

Wild Horses in a "European Wilderness"

In July 1999, eighteen Polish-bred semiwild horses were shipped to Latvia from the Netherlands and released onto a tract of overgrown meadowlands in a remote province on the Baltic Sea coast. The Konik horses had been brought to Latvia by a group of Dutch environmental consultants to help restore the Lake Pape coastal wetland ecosystem. For the Dutch team, this moment heralded a small but meaningful achievement in a larger battle to create "wilderness" in Europe, to carve out places where large wild animals could roam freely and natural processes function with minimal human interference, on a continent thoroughly subdued centuries ago by agricultural cultivation, industrial development, and human settlement. For the Dutch group's local partners at the recently established Latvian program office of WWF-International, the arrival of the horses stimulated a sorely needed national dialogue. Latvians needed to reconsider the significance of rural landscapes, the relationship between Latvians and nature, and the contours of rural development in the post-Soviet era. But for another major conservation organization, the Latvian Fund for Nature (LDF), the introduction of the Konik horses signified that "an enormous territory [had] been taken out" of the Latvian landscape: that the nationally iconic Latvian farmer had been driven off the land, to no good purpose, by "animals of fairly bizarre genetic origins."

The wild horses came to Latvia as part of an internationally funded project to promote conservation and sustainable development along the Baltic Sea coast. At Lake Pape, this aid initiative transformed an isolated, sleepy village into an active site for exploring the local meanings of sustainable rural development. In the competing responses of WWF, the LDF, and local defenders of the ethnoscape, we can again see the structuring role of the agrarian-internationalist discursive divide.

From Central Planning to "Stakeholder Democracy"

In the turbulent years of the fin de siècle, during the First Awakening of anticolonial Latvian national consciousness, Riga was the multicultural hub of the Baltikum. But Riga was not the only site for the transmission of Western ideas. As

Arveds Švābe notes, the growing port city of Libau (Liepāja), in the Kurland region, was "the gateway through which illegal [socialist] literature from Germany entered Latvia," and some 20 percent of adherents to the social-democratic New Current in Latvian political culture received their schooling there.[1] As Latvia's physical link with the outside world, the Baltic Sea coast has always been a gateway to internationalism. But under Soviet rule the entire coastal region was sealed off as a militarized border zone. It is thus fitting that in the post-Soviet era the Kurzeme coast became the site of Latvia's first major foreign-sponsored environmental aid project.

In 1992, environment ministers from the Baltic littoral adopted the Baltic Sea Joint Comprehensive Environmental Action Program, which stressed the need for pollution control and better management of coastal lagoons and wetlands. In Latvia, one outcome of this program was the multilateral Liepāja Environmental Project, sponsored by the World Bank. Approved in 1994 and launched in early 1995, the Liepāja Project sought to reduce pollution discharges into the Baltic Sea to protect biodiversity, "to re-integrate Liepāja and its region into the Latvian national fabric after more than forty years as a closed military city under Soviet occupation ringed by a countryside full of high security installations, and to develop a diversified economy in the region to replace the former 'military economy.'"[2] The project included four management planning demonstrations: an Integrated Coastal Zone Management (ICZM) Plan for the entire Kurzeme coast, the Kurzeme Ecotourism Development Plan, and local management plans for two selected sites, Lake Pape and the village of Jūrkalne.

Perhaps the most radical aspect of the Liepāja Project, in the post-Soviet context, was to introduce into Latvia Western methods of participatory planning. While few can agree on what sustainable development means in practice, all of its proponents agree (at least in principle) that in any given instance, this question can be answered only through the democratic participation of all parties involved. The UN's Brundtland Commission, which produced the seminal report on sustainable development, emphasized consensus building among diverse interests as crucial to solving global problems and hoped that such consensus could be reached through "dialogue, as if sustainable forms of development would best emerge from education, enlightenment and information. Such emphases led to a new rhetoric of partnership and stakeholder democracy."[3] In Latvia, this new rhetoric was introduced through ICZM, "a relatively new planning instrument based on full integration of policies and legislation on economic and socio-economic development, environmental protection and nature conservation" and on "the active involvement of all levels (e.g., local stakeholders, communities, regional/governmental authorities, interest groups)."[4] Each of the demonstration activities was overseen by a steering group comprising central and regional staff from the environment ministry, local and district authorities, scientists, and local community organizations.

The participatory approach represented a dramatic shift from the hierarchical, top-down planning style characteristic of the Soviet command-administrative economy. In addition to new substantive ideas about sustainable development, Latvian participants in the ICZM process appropriated new decision-making techniques and their accompanying buzzwords: consultation, feedback, public hearings, and SWOT analysis.[5] The newness of this approach was illustrated by the voluminous glossary appended to an early planning document, which included definitions not only of foreign terms (Delphi analysis, logframe, time-critical path), but also of familiar words and phrases (goal, strategy, beneficiaries) that had acquired entirely new meanings.[6]

Appropriately, the last item in the glossary was *vision*. For the first time since before Soviet rule, Latvians were permitted to articulate a development vision for their region and local community, to debate it among themselves, and to craft strategies for its realization. Through the rationalizing technologies of Western-style strategic planning, Latvian participants were encouraged to analyze assumptions previously taken for granted. As Maruta Kaminska, a local environmental official and ICZM participant, observed:

> This international aid is very essential . . . and not just in the financial sense. . . . The many people who work with these projects, they rise to a new level, and they don't have problems any more. . . . Sure, we already know how to say very generally that nature here is untouched in many places and so forth, but all that has to be backed up with specific research materials—how much, where, the sites with all the cartographic materials. We didn't have opportunities like that before. . . . [These projects] train people, because if someone has participated even slightly, he already thinks completely differently, and his perceptions are different too.[7]

In short, the ICZM process impelled Latvians, from local activists to national ministry officials, to think strategically about the goals and methods of rural development, activating dialogue about values to be defended, opportunities to be exploited, and threats to be fended off.

The Lake Pape Project

Located in the remote rural township of Rucava, hugging the Lithuanian border in the southwestern corner of Latvia, Lake Pape made an ideal ICZM demonstration site, combining poor socioeconomic indicators—economic decline, sparse settlement, very low and declining population—with high nature values. The relatively small project territory (273 square kilometers) contains a diverse range of ecosystems, including a large coastal lagoon (Lake Pape) with coastal wetlands and reedlands, a sandy beach with shifting dunes, a high peat bog (Nida Bog), wet meadows, and the alluvial forest and streams of the Sventāja River valley. Wet birch and alder forests, along with pine and spruce forests on higher ground, provide habitat for many large mammals, including moose, elk, deer, wild boar, beaver, otter, raccoon dog, and wolf. Lake Pape is also a major bottleneck on the Baltic portion of the East African–European–Arctic migratory flyway. As a "critical part of a system of coastal wildlife habitats which includes the Kursiu Lagoon to the south in Lithuania and Matsalu Bay to the north in Estonia," the area has been "designated as an internationally significant area of nature conservation concern by HELCOM [the Baltic Marine Environment Protection Commission], IUCN and WWF." [8] The Latvian government has proposed Lake Pape and Nida Bog as Ramsar sites, to be protected under the Convention on Wetlands of International Importance Especially as Waterfowl Habitat, and as protected zones within the EU's Natura 2000 network.

While human settlement of the Lake Pape area dates back to the Stone Age, agriculture and cattle herding appeared in the seventeenth century and were concentrated in the higher land around the modern village of Rucava. "The influence of low-intensity agricultural activity on the landscape was minimal but essential; grazing by domestic animals and annual mowing checked the natural growth of vegetation," thereby maintaining a diverse range of habitats, especially for water birds.[9] Reed cutting in the lake fostered a favorable environment for insects and fish, which in turn allowed otters to thrive. Extensive farming and grazing practices left the wetland ecosystem relatively unscathed, although there were early attempts to regulate water levels for flood control and agricultural purposes. The first artificial dam was built in 1830 and the first sluices in 1847; modern sluice gates were erected in 1923, shortly after the establishment of the independent Latvian state.

Soviet rule brought agricultural collectivization and intensification, including heavy use of chemicals and drainage of wet meadows and forests, thereby ending the synergy between farming and biodiversity. To regulate water levels for drainage and reed cutting, the sluice gates were upgraded in 1950, and Lake Pape's two small contributory rivers, the Paurupe and Līgupe, were channelized to bypass the lake and discharge their water directly into the sea. The ecological consequences of these actions were profound: the water table dropped drastically, the meadows dried out, the lake silted up, reed cover increased considerably, and water birds

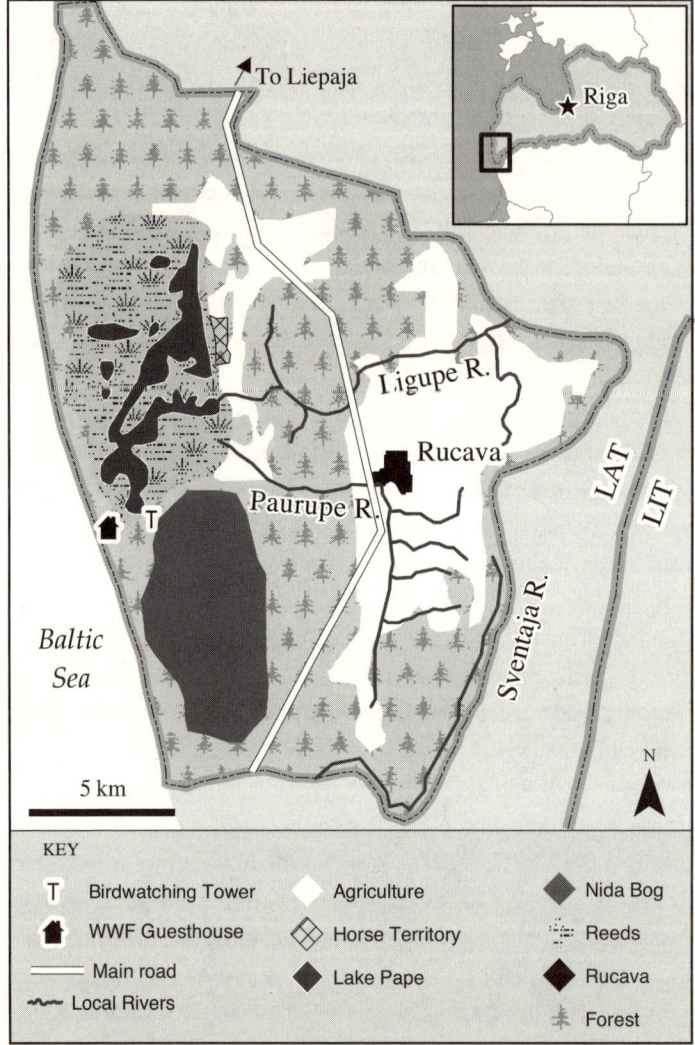

Figure 9. Lake Pape ecosystem. Map: Robert D. Lopez, adapted from WWF-Latvia.

declined because of loss of habitat and food. Other ecological impacts from the Soviet era included peat excavation on Nida Bog and bombing in the military ranges. The onerous restrictions of the closed military zone meant that nature was left relatively untouched in some areas, with ambivalent ecological consequences: while some areas benefited, biological diversity in meadows, for example, suffered from the absence of grazing pressure. Portions of the territory were placed under protection during the Soviet period, including the Lake Pape Ornithological Reserve, the Kalniški Complex Nature Reserve, and several smaller geological and botanical reserves.

In 1940 the Rucava area was relatively prosperous, but this status reversed dramatically during Soviet rule and the post-Soviet transition. The human population was decimated by deaths and emigration during World War II, including the wholesale elimination of the Jewish community, and then by deportations attending collectivization. From a high of 5,000 in the 1930s, the population had dwindled to 1,890 by the 1990s, or seven persons per square kilometer. Collective farms were virtually the only source of employment during the Soviet period, but these were liquidated after independence and their inventory snatched up by the well-connected few. The dual impact of collectivization and militarization under the Soviets left the area "short of people with higher education, business capacity or innovative experiences."[10] The close proximity of Lithuania, with its cheap, nationally subsidized agricultural products, and poor transportation links to Liepāja and other Latvian population centers, made it almost impossible to market local produce. Moreover, the soil was poor and there were no local processing facilities. Most local residents had reverted to subsistence farming. As the lack of economic opportunity drove more and more young people out of the area, the population was aging rapidly.

Soviet rule profoundly transformed the cultural landscape as well as the local ecology and economy. Those farming families not deported were moved from farmsteads and scattered hamlets to village centers. Many were housed in shoddily built, unattractive low-rise apartment blocks, which remain today as decrepit testimonials to the Soviet vision of proletarianizing the countryside. The fishing hamlets of the pre-Soviet era were emptied, too, as coastal fishing was prohibited by military restrictions. The countryside around Lake Pape was thus a classic example of the landscape of abandonment: overgrown fields, deserted and crumbling farmsteads, the skeletal remains of kolkhoz outbuildings, long ago stripped of any valuable materials.

Such were the challenges faced by the Lake Pape Project, which set out to "develop and implement a community-based, ecologically sustainable, environmental management plan and . . . to use an approach which could serve as a model for replication in Latvia as well as within the Baltic Sea region."[11] The project was funded by the Danish Ministry of Environment and Energy and implemented jointly by WWF-Denmark and WWF-Latvia. The planning team concluded that without a fundamental reversal in nature management practices, within twenty years Lake Pape would likely be completely overgrown with reeds, leading to pro-

Figure 10. Soviet apartment bloc, Rucava. Photo: Katrina Z. S. Schwartz.

found loss of biodiversity. They articulated two imperatives for preventing this loss: "(1) maintain and improve the biodiversity of the area, with emphasis on restoration of the Lake Pape Ecosystem; and (2) [restore] the local economy and business structure through sustainable development based on local characteristics and [strengths]."[12]

Some measures were implemented during the first phase of the project. With the deintensification of agriculture after independence, the water-regulating system of channels, polders (drained, sub–sea level agricultural fields), pumps, and sluice gates had fallen into disuse and disrepair, thus creating political space to press for restoring natural processes in the lake. Bowing to local demand, the steering group agreed to rebuild the sluice gates, but at the same time it won approval to dechannelize the Paurupe and Līgupe rivers, which partially restored the natural flow of water into and out of Lake Pape. Modern reed-cutting machinery was purchased, enabling local residents to earn income from the export of reeds (now popular as a roofing material in Denmark and Germany) while also helping to increase the lake's open water. Measures were taken to promote ecotourism as a key component of the area's future economic development, including construction of a bird-watching tower on the lake and some modest visitor facilities. Project staff persuaded two local families to open bed-and-breakfast operations that initially catered largely to project-related visitors from Riga and abroad. Toward the goal of building local capacity for "diversification of the rural economy," a credit union was established and adult education courses were offered, and widely attended, in tourism, foreign languages, business education, and basic computer skills.

Toward a "Common Vision"?

The Lake Pape Project, according to its implementers, was "very process-oriented, focusing intensively on the achievement of a common language on how to analyze the situation . . . and of a common understanding of the problems and solutions."[13] The "common language" to be fostered was, of course, the Western language of sustainability. Latvians were to learn to see land through the lens of nature value; biodiversity was to be embraced as the mechanism for restoring value to a devalued landscape: "Thus, a main task of the project has been to direct the local community's view of their natural environment and its values into one of an asset."[14] The principal targets of this exercise were local district and township

Figure 11. The landscape of abandonment. Photo: Katrina Z. S. Schwartz.

authorities, who in Latvia are critical actors in zoning and rural development decision making, owing to the central state administration's weak capacity in this
sphere.

By all accounts, the Lake Pape Project and its counterpart in Jūrkalne were
enormously successful in raising consciousness at the local government level.
Given the passivity and inertia of local governments in the post-Soviet era, the
leaders of both townships stand out as vivid exceptions. According to Valts Vilnītis, who chaired the joint steering group for the two projects, "Their attitudes
changed enormously, they became active, got their people involved."[15] Mayors of
both towns pragmatically accepted that, given the remoteness and poor soils of
their regions, as well as current market conditions and state agricultural policies,
small-scale farming would not be the backbone of local rural development. "We
don't have to cling to the Latvian farming way of life as if to an Indian holy cow,"
asserted Māris Dadzis, mayor of Jūrkalne.[16]

Yet both mayors argued strenuously that while keeping people in agriculture
should not be a primary goal of rural development, keeping people living and
working in the countryside certainly should. As Dadzis put it, "How many can
cram into Riga?" Through the ICZM experience, both mayors had mastered the
art of preparing Western-style funding proposals and were using it to promote diversified local development. Rucava's mayor, Jānis Veits, in particular, had become renowned for his grant-writing prowess and was the subject of a lengthy
profile in the national daily, Diena, under the headline "Annual Budget Doubled
Through Grants." "Rucavites' belief in grants and, more important, in Rucava's
value began with the Pape-Jūrkalne project of 1993," observed the Diena reporter.
As the township's project director recalled: "The Danes came here, and we ourselves started to feel that we could do something here."[17] Between 1993 and 1999,
local officials and private citizens submitted some thirty grant proposals to international donors (WWF, EU Phare, the Soros Foundation, the government of Sweden) and to Latvian agencies (the state employment agency, the environment ministry, the economics ministry's Regional Fund, the Environmental Protection
Fund, the Cultural Capital Fund), for a total of at least 40,000 lats. All but eight received funding. In addition to some modest investments in infrastructure, such as
building an open-air market in the town square, grant monies were used to create
several dozen jobs, both temporary and permanent. Seeking to involve a wider
network of people in promoting local development, Mayor Veits organized the
first worldwide reunion of Rucavites in the summer of 1999, complete with a conference entitled "Rucava Yesterday, Today, and Tomorrow." In his address to the

gathering, Veits stressed the growing importance of seeking grants and the diminishing role of farming. "Agriculture," he declared, "is a rich man's hobby."[18]

Both mayors enthusiastically embraced the notion of using nature as an asset for achieving economic revitalization. The local government was developing small-scale ecotourism facilities in Jūrkalne, with its wide beaches and high, clifflike dunes. "We don't want to become another Jūrmala," Dadzis assured me, referring to the overcrowded and overbuilt resort community just west of Riga, whose beaches are scarred by the monstrous high-rise hotels and spas of the Soviet era. Rucava's Veits was hopeful about the development potential of nontraditional enterprises such as "cranberry growing [in the bogs], reed cutting, ecotourism on the seacoast and in the Sventāja River valley, boating in the lake and the sea for tourism and fishing."[19] Nevertheless, he predicted that small grant-funded economic diversification projects alone would not be able to generate long-term employment for most local residents:

> The most painful thing is that with all our efforts at attracting funding and implementing projects, it's clear that through the new development plans we might be able to involve only around half of the township's residents. We won't all be ecotourism and rural tourism providers, but only some 10–30 percent of the population. What will the rest do, the farmers? We can't all switch to growing currants and strawberries. This question we haven't answered even theoretically. In this remote corner we're driven into a wedge—the sea to one side, the Republic of Lithuania to the other, and there's nowhere to go. Jobs are not being created at the national level. The youth are leaving for the cities. The countryside is at risk of becoming an old people's nursing home.[20]

Lacking "even theoretically" a broad-based program for sustainable rural development, Veits appeared willing to embrace any and all prospects. While he enthusiastically endorsed WWF's efforts at nature restoration and ecotourism, he and his staff also entertained a range of other possibilities considerably at odds with a development scenario favoring biodiversity: constructing a golf course or airstrip, excavating peat for export, renewing and maintaining drainage systems and polders.[21] His activism thus appeared to be driven less by an overarching vision than by the availability of funding opportunities. Indeed, as Veits himself stated frankly to the Rucava conference, "a project to us means additional money."[22]

In short, while participation in the ICZM process had generated exceptionally active interest in how to pursue sustainable rural development in the Lake Pape area, the local community had yet to coalesce around a definitive response to this challenge. Two alternative strategies were being promoted by local defenders of the Heimat on the one hand, and the internationalist and radically postagrarian staff of WWF-Latvia on the other.

"A Museum of Ancient Latvianness"

Independently of the international activity around the ICZM process, local school-teachers in Rucava had for some time been quietly engaging in homeland studies, now known as regional studies (novadpētniecība), or the historical exploration of the local cultural landscape. Digging through archives and questioning survivors of the pre-Soviet era, they chronicled the area's ancient history and archeology and had unearthed and recorded a wealth of microtoponyms—traditional names of natural objects, homesteads, vanished hamlets, and so on, along with their etymological origins. In 1994, local enthusiasts formed the nongovernmental Rucava Nature Fund (RNF). The fund's name clearly signified nature as ethnoscape. As Ināra Rūce, a biology teacher and chair of the RNF, put it: "We are not a pure nature fund; we are also concerned with the cultural environment."[23]

Not herself a native Rucavite, Rūce was struck when she first came to the area by the unfamiliar plant life, which differs from that of inland Latvia owing to the mild coastal climate. But she was also amazed that "there are so many ethnographic values here, in daily life." Indeed, if Lake Pape had long been known to Latvian ornithologists as one of the country's migratory hot spots, then Rucava was also renowned as a sort of living ethnographic museum. Already in the late 1800s, a visiting Latvian cultural figure described Rucava as "a museum of ancient Latvianness."[24] A travel brochure produced in 1941 by the national open-air ethnographic museum observed: "In this isolated, peculiar place, ethnographic phenomena that have long since vanished elsewhere, no longer surviving even in memories in other Kurland regions, have been preserved as if in a kind of museum. Here it is all still alive and vivid."[25]

Throughout the dramatic upheavals of the twentieth century, the area's geographical remoteness, reinforced in the Soviet period by the compulsory isolation of the closed military regime, helped preserve the ethnographic fabric of life in Rucava "as if in a pickle jar," in the words of Gunta Timbra, town librarian and RNF founding member.[26] Left largely to their own devices, with restricted in-migration from the outside world, Rucavites reportedly continued to wear their region's traditional folk costumes on special occasions as late as the 1970s. RNF members were quick to point out that many local folk traditions were still actively maintained. "The first few years [of living in Rucava] I walked around completely amazed—grannies come to the store wearing the classic Rucava knitted mittens," Rūce told me. "If someone knits socks, then they knit them [in the traditional Rucava style], not from a pattern in some magazine or from some store, or from another Latvian region."

The ethnoscape thus weighed equally with the natural environment in the brief Rucava almanac published by the RNF, which celebrated the historically synergistic relationship between farming and ecology in the classic spirit that equates Latvianness with closeness to nature:

> Rucava township is wealthy not only with natural diversity. Anyone who has been there . . . immediately notices how nature itself is inextricably bound up with the human dwellers, with their harmonious environmental sense, with their employment and way of life, their traditions of building and everyday life. Precisely this harmoniously created, and still preserved, human living environment is one of the most valuable elements (perhaps even as valuable as the natural diversity) of the heritage of Rucava township.[27]

Along with detailed descriptions of the local geology and ecology, the Rucava almanac included chapters on cultural history, with descriptions of ancient ritual groves and stones, folk arts and crafts, and vernacular architecture, asserting:

> The coastal landscape is unimaginable without the fishing hamlets and fishermen's homesteads. . . . The ancient architectural traditions in the hamlets are also connected to nature: natural materials—wood, reeds, straw, fieldstones—are used, and the buildings blend well with the surrounding landscape. . . . Even the transformations of the late twentieth century barely affected the township's coastal area, and thus today the late nineteenth-century and early twentieth-century fishermen's and farmers' homesteads can still be seen, harmonizing well with the area's primordial, relatively untouched nature.[28]

The almanac pointed out that the fishing hamlets had been designated as architectural monuments.

This ethnographic heritage was featured prominently at the Rucavite reunion in 1999, with folk music performances, the display and sale of local crafts at the newly built marketplace, and a restored historic dwelling opened in the town center. For RNF members and many other local residents, homeland studies represented not simply a hobby or a fond memory of "Rucava Yesterday," but also a vision of "Rucava Tomorrow." Thus the librarian Timbra maintained that local development should "spring from cultural history." In her dream of Rucava's future, traditional local crafts would be practiced, and residents would have Rucava folk costumes at home and "not be shy about wearing them in public." In time, "the ancient values will again be valued."[29] In the fall of 1998, Rucava and the neighboring Lithuanian town of Kretinga received funding from the EU's CREDO cross-border development program to prepare a management plan for the Sventāja River valley, which marks the Latvian-Lithuanian border. The focus of the plan, it was agreed, would be tourism development, and the Latvian-side project leader proposed promoting Rucava as "a place which speaks to people through people." Rucavites attending an early project meeting enthusiastically endorsed the idea of basing tourism on local traditions and everyday life, offering suggestions such as posting plaques on houses bearing family names and histories, holding contests for the best-tended farmstead and offering tourists the opportunity to try their hand at farm labor.[30]

For some local enthusiasts, contact with internationality through the Lake Pape Project had leavened the cultural landscape orientation with a heightened awareness of the area's nature values. As Gedimins Salmiņš, a physical education teacher and amateur scholar of local history and archeology, recalled:

> We started to recognize some ten years ago that here in the Rucava area the nature is unique, that there is much preserved here that doesn't exist elsewhere. The ornithological station [established in 1977], where they do bird banding, could have drawn attention earlier, but right there beside the banding station they had military drills. . . . For me personally, I was somehow suddenly impressed by the first reports that now, after the collapse of the Soviet regime, scientists from Sweden had shown up here and were interested in the things we ourselves had passed by or trampled on and were ignorant of.[31]

Ethnographic homeland studies, in other words, could be compatible with capitalizing on internationally recognized ecological values. Thanks to the historical synergy between nature and the traditional agrarian lifestyle, Salmiņš and Rūce could imagine the Lake Pape region simultaneously as a museum of local culture and as a globally significant migratory corridor for birds. A fairly close collaborative relationship developed between the RNF and WWF-Latvia, with the former providing local personnel and the latter moral and material support in the hopes of grooming the RNF to become the primary local agent of its conservation and development vision.

But many other local residents derided the sort of untrammeled nature that might interest Swedish scientists as worthless "jungle," unfit for human enjoyment unless redeemed through cultivation. A participant in the Sventāja River Project meeting wondered what would be done about the untended, tick-infested riverbanks, which in her mind were "nothing but an impassable jungle." The tourism plan should focus on "acquainting tourists with our culture and traditions, instead of that undergrowth," another concurred: "There's no fun in tramping through those brambles!"[32] RNF member Rūce countered by pointing out that the area had "botanical riches that make Riga botanists' jaws drop," and that among tourists, "some will want to see ethnography, and others—a jungle."

Rucava's many decades in the "pickle jar" had not only kept grannies outfitted in traditional socks, but also fostered an intensely local sense of place. In the winter of 1999, I helped WWF administer an opinion survey of area residents, aimed at assessing attitudes toward WWF projects past, present, and future. To the question, "Is the nature here special or worthy of protection?" almost all of the fifteen or twenty respondents I interviewed answered in the negative, although when I mentioned Lake Pape or Nida Bog, they invariably responded: "Oh, yes, well, over there in Pape . . ." Many of these lifelong residents of the half dozen scattered hamlets of Rucava township had never traveled the five or ten miles to Pape or Nida or the Sventāja River valley. Military restrictions had inhibited travel in the Soviet period, and since independence, poverty had the same effect, as gasoline had to be spared for more practical uses.[33]

Poor soil and market conditions notwithstanding, the developmental vision of agrarian nationalists had many adherents in Rucava. According to the librarian Timbra, "It will be disastrous if the fundamental employment in the future will not be agriculture." In her view, tourism could at best be only a supplementary activity.[34] Voldemārs Timbra, head of the local farmers' association, was frankly scornful of the Western approach to sustainable rural development. "These international projects are ready to support anything—horses, plants, et cetera," he declared; "anything except what is really needed—one solid demonstration farm!"[35] A retired couple whom I interviewed for the WWF survey passionately denounced the abandonment of the agrarian landscape, and particularly the overgrown drainage ditches and flooded fields. The wife recalled how her father and grandfather had tended their thirty hectares during the interwar years. They had no machinery, "only their hands and horses, but as soon as any little shrub grew up in the drainage ditch, they removed it." In the Rucava section of the Kurzeme ICZM plan (prepared separately from the WWF-led Lake Pape Management Plan), one of the chief goals identified was "preservation of the existing area of agricultural land," and proposed activities included restoring abandoned polders and renewing peat extraction on Nida Bog—activities aimed at preserving not globally vital East European wetlands, but an agrarian Latvia.[36]

In short, despite the ICZM invasion, many—indeed, probably most—Rucavites continued to imagine local development along traditional agrarian lines. Even among those who embraced the new notion of economic diversification, most envisioned a local tourism industry based less on wetlands and migratory flyways than on the cultivated landscape, or Heimat. If the government of interwar Latvia had subsidized the creation of an open-air ethnographic museum to entrench and reify the agrarian conception of Latvian identity, then Rucava's post-Soviet Heimat enthusiasts hoped to preserve that identity by "selling" themselves as a living ethnographic museum. As we have seen, the European paradigm for sustainable rural development does support linking traditional farming lifestyles with new postproductivist values. In its analysis of protected nature areas and sustainable rural development, for example, IUCN endorses the promotion of heritage tourism through activities such as "encouraging rural communities to develop local museums of rural life, or other ways of celebrating their relationship with nature."[37] Protection of the ethnoscape, in other words, can be a legitimate component of a sustainable rural development strategy in the new Europe.

However, given the enormous administrative and budgetary hurdles facing the EU's strategy of low-impact farming and green recoupling (subsidizing farmers for environmental protection activities), and given the absence of an agrarian-oriented regime like that of Latvia in the 1920s and 1930s, Rucavites faced the thorny question of who would pay to support the agrarian lifestyle or its marketing through heritage tourism. Even in affluent Western Europe, according to Brian Ilbery and Ian Bowler, "the idea that farming is needed everywhere to maintain valued environments will need to be re-examined, with the creation of wilderness areas becoming a distinct possibility. This will be controversial with many member states, particularly where the cultural desirability of a 'landscape with figures' is more firmly entrenched."[38] In Rucava at the end of the 1990s, defenders of the agrarian ethnoscape had not yet begun to grapple with the practical dimensions of how to make economic development "spring from cultural history." Pragmatists like Mayor Veits feared that neither farming nor grant monies would be adequate to maintain the traditional "landscape with figures," but they had not yet articulated a compelling alternative. Meanwhile, a radically postagrarian alternative, prescribing a very different role in nature for humans, was being promoted by another set of ICZM participants, WWF-Latvia and its partners from the iconoclastic world of Dutch "nature development."

"Making Nature" in a Postproductivist Europe

With the EU's highest population density and most intensive land use, the Netherlands leads the union both in landscape degradation and in efforts to reverse this degradation. "Force of circumstances—dense populations of both people and animals on a small area, a high water table, and proximity to 'Europe's dustbin and sewer,' the North Sea—makes the Dutch the most environmentally conscious nation in western Europe."[39] This consciousness has yielded not only Europe's toughest pollution laws, but also the most active engagement with the relatively new applied science of restoration ecology. Pioneered in the United States in the 1930s and increasingly popular since the mid-1970s, restoration seeks to renew the functioning of natural processes in human-transformed environments.[40] In the Netherlands, the shrinking and fragmentation of nature reserves has inspired conservationists to adopt an aggressive restoration strategy, which they call "nature development," aimed at "the creation of entirely new habitats."[41]

This practical restoration work is underpinned theoretically by a radical revision of Europe's ecological history. According to the mainstream theory, the preagricultural European landscape was a dense, unbroken forest, in which open land developed only as a result of human intervention in the form of farming. Without continued grazing by domestic livestock, in this view, open spaces would become entirely reforested, resulting in a loss of many existing habitats. Against this orthodoxy, Dutch scientists have argued that preagricultural Europe was in fact a shifting, parklike mosaic of grasslands, scrub, groves, and solitary trees. Grazing by wild herbivores was the critical factor in creating open areas and maintaining vegetative diversity in this mosaic landscape. Not only were the familiar moose, deer, elk, and beaver present in much greater numbers than today, but so, too, were the now-extinct tarpan (wild horse) and aurochs (wild bovine) and the near-extinct wisent (European bison). Each species of herbivore occupies its own niche in the food chain: deer, moose, and elk eat tree and shrub leaves and bark, for example, while horses and cattle largely eat grasses and herbaceous plants.[42] According to the Dutch researchers and like-minded colleagues, only the presence of the full hierarchy of herbivores can maintain a diverse and viable mosaic ecosystem. In much of continental Europe today, however, the ecological role of large grazers has been largely usurped by humans through domestication, hunting, and intensive forestry.

The Dutch scientists draw the heretical conclusion from their findings that "species diversity in Europe is not a result of the introduction of farming," but rather a result of natural processes.[43] Mainstream European conservationists seek to protect current species by supporting traditional, low-impact farming practices. The Dutch revisionists argue instead that this is a very costly way to achieve conservation goals and that while it does help preserve some species, it endangers others through fragmentation of habitat and maintains an artificially static landscape. A more economically and ecologically sustainable approach, they argue, is to restore natural processes by intensifying agricultural production on a smaller land area, taking marginal lands out of production altogether, and establishing "a network of large nature reserves with a minimum of human interferences."[44]

Since the mid-1980s, Dutch conservationists have been testing their theories by pioneering efforts to introduce into the wild formerly domesticated large grazing mammals. Several herds of the small Polish horse, the Konik polski, the closest surviving descendant of the extinct tarpan, have been released in nature development areas. Similar experiments have been made with several bovine species with primitive traits: the hornless Galloway, the long-horned Scottish Highlander, and the Heck bovine.[45] Dutch nature developers have also joined WWF-International in spearheading broader efforts to restore wild mammal populations. The Large Herbivore and Large Carnivore Initiatives, launched in the mid-1990s, seek "to maintain and restore, in coexistence with people, viable populations of large carnivores [and herbivores] as an integral part of ecosystems and landscapes" across the Eurasian continent.[46] WWF and its partners are working to increase the range and distribution of wolf, brown bear, and lynx throughout Eurasia, to restore camels in Mongolia, and to return "Europe's largest wild herbivore, the European bison, to its natural range" in Russia.[47]

The cultural significance of the iconoclastic Dutch nature development vision is clear when compared to its American counterpart. The most prominent strand of American restoration ecology is something of a quasi-spiritual movement, devoted to renewing not only natural processes but also human beings' relationship with nature.[48] Reacting against both the utilitarian discourse of rational exploitation and the preservationist discourse of nature as pristine, sublime wilderness (and therefore separate from and untouched by humans), restoration ecology reconceives the role of human beings as neither masters nor intruders, but rather as gardeners. In this view, people can enrich both themselves and nature by "tending" wild nature. "Arcadia," declares the philosopher of restorationism Frederick Turner, "is a place where human beings cooperate with nature to produce a richness of ecological variety that would not otherwise exist."[49] The labor of ecological restoration provides a way for exiled humanity to return to Eden and reconnect with our own forgotten history, "the history of our interaction with a particular landscape, and the deeper history of the general relationship of our species with the rest of nature."[50] In imagining nature as a reservoir of human history and a site for the human labor of cultivation, and in imagining that labor as constituting Americans' authentic identity through "full citizenship in the biotic community,"[51] restoration ecology can be seen as an American form of protecting the ethnoscape.

As a reaction against this dualistic attitude—nature as exploitable resource and as untouchable wilderness—advocates of restoration in the United States have shifted toward what in Europe has long been the dominant understanding of the relationship between human beings and the natural world. While Europeans have constructed nature as Edenic wilderness in their colonial domains, they have rarely celebrated their native terrain in those terms, but rather as cultivated landscapes for human use. European national parks are typically not "pastoral paradises preserved from evidence of human labor, but working agrarian landscapes," writes Kenneth Olwig. Unlike their American counterparts, they tend to be conserved precisely because of their evidence of ancient habitation and stewardship, and it is widely recognized that the landscape must continue to be worked by the local community if it is to exist."[52] At France's Parc de Paysan, for example, "as in other French national parks, the maintenance of a population of sufficient size to support the traditional landscape is central to its philosophy," and park authorities strive to "preserve the right balance between fauna, flora, man's cultural heritage and the economic prosperity of the region."[53] "No national park in England and Wales can ever approach the national park ideal as exemplified by Yosemite," notes a British commentator; "for us agriculture and forestry, in particular, will have to take their place."[54] As David Lowenthal observes, "European national icons today stress intimate, humanized chocolate-box scenes—a figured landscape, whose traces of cultural heritage are now often embellished. Photographers' props in Sweden enrich the classic Dalecarlian Lakeland scene: a collapsible wooden fence, a model in folk costume, a replica birch-bark horn, a couple of goats."[55]

Thus in the Netherlands, where the last bits of "wild" nature have long ago been brought under human management and cultivation, there has been no need to remember or revalorize man's place in nature, and restoration has played a very different role than in America. If the ultimate goal of both U.S. and Dutch restoration could be defined as "making nature,"[56] then American restorationists emphasize the "making," while the Dutch emphasize the "nature." Because Europeans have long understood nature as a cultivated garden, the Dutch employ restoration to reimagine nature as wilderness. And because man's historic role in the landscape has never been forgotten, they articulate a new, more limited role of managing for natural processes and seek to reconnect with an earlier, almost prehuman natural history: the history of the preagricultural mosaic landscape of wild grazers.[57]

The Horse Project: "Restoring a European Wilderness"

As the first phase of the Lake Pape Project wound to a close, WWF-Latvia director Uģis Rotbergs solicited funding from WWF-Sweden for a feasibility study on restoration of the ecosystem through grazing of the wet meadows. Searching for a contractor to carry out the study, Rotbergs discovered the Ark Foundation, a Dutch consultancy involved in "nature development" by reintroducing wild herbivores. Rotbergs was intrigued, and despite the reservations of his Swedish donors, who had anticipated a more mainstream approach based on low-impact grazing of domestic livestock, he persuaded them to hire the Dutch team.

The Dutch feasibility study identified the Lake Pape territory, a large area with little human activity and a high diversity of fauna, as an ideal site for nature restoration:

> Large grazers such as elk, red deer [moose], roe [deer], beaver and wild boar are still present in low densities, and of the large predators, the lynx, fox and wolf still hunt. A large and relatively undisturbed high peat moor covers parts of the area. A relatively small number of people live around the lake. The soil is only marginally suitable for agriculture and forestry and many farms are unoccupied. Besides two partly inhabited fishing villages, there are some recreational houses. . . . This area is large enough to lodge all large indigenous mammals (both grazers and predators). The most important species are already living there and can, when protected, reach their natural densities.[58]

The Dutch team proposed completing the full pyramid of natural grazers by introducing first dedomesticated horses and eventually cattle and bison. These additions would make the Lake Pape area a nearly "complete" ecosystem; except for the brown bear, which requires a larger territory, all of the large mammals that dwelled in this region before the agricultural era would be present. The Pape area would thereby become unique in the European context. As the Dutch study noted, "A nature area where nearly all of Europe's fauna of large grazers and predators live is rare, even on a European scale. It may serve as a pilot area for similar projects in Europe where agriculture is withdrawing."[59]

The Ark Foundation donated eighteen semiwild Konik horses to WWF-Latvia. With additional funding from WWF-Netherlands and the Large Herbivore Initiative, in July 1999 the horses were released into a fenced 250-hectare territory around Lake Pape, leased by WWF from local landowners. The Ark Foundation provided a resident horse expert for the first months of the project. The small, hardy, sandy-gray horses share many primitive traits of their tarpan forebears that enable them to survive in the wild, as does their social structure.[60] Koniks live in harem groups comprising a number of mares and one leading stallion; mares pass down knowledge about coping with the local environment and predators from generation to generation, as well as when joining new harems.

The Ark team predicted that the Koniks would adapt to their new environment and would thrive and reproduce with little or no human assistance, belying local

Figure 12. Konik horses at Lake Pape. Photo: Katrina Z. S. Schwartz.

fears that the horses would succumb to starvation, thin ice, or poaching. Willem Overmars, project leader for the Dutch side, noted that project staff would have to work diligently to counteract the "local knowledge" about these animals, by teaching area residents to view the Koniks not like their domestic relatives but as wild animals that must be allowed to learn from experience.[61] The horses would reestablish grazing pressure on the meadows around the lake, thereby enhancing biodiversity at vastly less cost than the agrarian alternative of subsidizing small farmers to keep dairy cattle or other domestic livestock. Ecological conditions would be monitored at regular intervals to assess the impact on biodiversity. Eventually the fences would be removed and the horses would roam freely, coming into unmediated contact with predators and fully integrating into a natural system that "actually covers the whole of Latvia and Lithuania."[62] Ideally, the unfenced horses would someday encounter bears from Estonia, too. Mirroring the ongoing transformation of the European continent into a borderless polity, in other words, Lake Pape would become part of a borderless state of nature.

The introduction of Konik horses was intended as the first step toward creating a larger nature area. In collaboration with the Rucava Nature Fund and local authorities, WWF-Latvia was working to establish a Lake Pape Nature Park, which would bring the entire greater lake ecosystem under the authority of a newly established, as yet informal Lake Pape Management Board.[63] Ultimately the Lake Pape Nature Park would be transformed into what the Ark team called a "new wilderness": "a large unbroken area where natural processes can take place with the least possible disturbances."[64] Human activities—mowing, reed cultivation, manipulation of water levels—were to be kept to a minimum in the park territory. Eventually the sluice gates controlling Lake Pape's tidal flows would be eliminated. Along the Baltic shoreline, too, human efforts at stabilizing and controlling nature would be halted, giving free rein to nature's constant flux and dynamism. Reversing the generations-old, heroic Latvian tradition of "battling the shifting sand dune," the pine groves planted decades ago to combat erosion would be cut to restore the "very characteristic ecosystems which have become very rare along the coasts of the North Sea and the Baltic."[65] The Baltic shoreline would creep slowly inward; winds would once again blow drifts of sand about, forming small dune lakes and covering whole forests. Nida Bog would be protected from peat mining and other human incursions. As the landlocked foil to Lake Pape's tidal dynamics and habitat for moose and cranes, the bog is an essential component of the "complete" ecosystem. Agriculture would be banned from the lowlands around Lake Pape. Commercial fishing, which disrupts the marshland food chain, would also "die out with the last fisherman." The landscape would be shaped not by human cultivation, in other words, but by the natural disturbances of erosion, flooding, and herbivory.

WWF-Latvia: Globalizing the Ethnoscape on Behalf of the Nation

For the Dutch team, Lake Pape was interesting primarily because it possessed biodiversity resources unique in the European context, and the horse project was important insofar as it contributed to continent-wide initiatives in biodiversity protection and nature restoration. Hence, the subtitle of the feasibility study: "Restoring a European Wilderness." To make the landscape more wild or natural was also to Europeanize or globalize it, replacing the local knowledge of agrarian homeland studies with the extralocal values ascribed by Western conservation policy, ecological science, and demand for ecotourism. The horses of Lake Pape richly symbolized this globalization of the ethnoscape: Polish horses imported from the Netherlands by a global conservation organization to replace agrarian cultivation and restore a "European" ecosystem, toward the ultimate goal of creating a borderless "new wilderness."

Project planners hoped that the wild horses would not only provide a valuable ecological service as grazers, but also help attract tourists. For the Dutch consultants, developing ecotourism at Lake Pape, like nature development, was important first and foremost because it met a European need:

> The urbanized society in Europe feels the need for places where wilderness can be felt, where man-made regulations are held aloof, where rivers take their own course, and where animals live their own lives within their own social orders. . . . Apparently there is a spiritual need to get away from the artificial every-day routine and to experience its contrast: pure nature with its own order.[66]

Along with the area's ecological riches, they noted, Lake Pape's selling points included low wages and prices by European standards and the geographical proximity of "Germany and Sweden, where many of the potential visitors come from." Tourism promotion efforts should therefore focus on attracting Europeans who seek "adventure in an area shaped by natural processes." (The consultants mentioned tourists from Latvia and Lithuania very briefly, almost as an afterthought.) The West European "adventurous tourist" envisioned by the Dutch project planners would, most likely, not be attracted primarily by the Latvian ethnoscape.

Like their Western partners, Uģis Rotbergs and his staff at WWF-Latvia lauded the horse project as an advance for global biodiversity protection, and they realized that Latvia's internationally recognized nature values were WWF's strongest selling points when "shopping" for donor funds. But unlike the Eurocentric Dutch team, Rotbergs wanted to give priority to attracting Latvian tourists. The horse project interested him most not as a recreational resource for jaded Westerners, but as a vehicle for transforming Latvian consciousness. To this end, WWF-Latvia was seeking funding for an ambitious third-generation project in the greater Rucava area, aimed at building knowledge and capacity among local and regional authorities and residents to promote sustainable rural development and nature protection within the EU approximation process. The project included the following components: (1) developing the Lake Pape Nature Park, complete with visitor facilities and an ecotourism office; (2) education and capacity-building in sustainable farming and forestry, small-business management, and administration of EU accession processes; and (3) small-scale demonstration activities in nature restoration, sustainable farming and forestry, energy efficiency, and tourism. In later phases, the project was to be "marketed" throughout Latvia to disseminate know-how and training for these objectives and to support lobbying for national development policies.

In short, WWF-Latvia endorsed the Dutch consultants' globalizing agenda, but did so first to serve a national purpose: namely, to change the way Latvians thought about nature and development. Indeed, Rotbergs aptly illustrates the fact that the outward-looking discourse (although I call it liberal *internationalism*) is still very much a discourse *within* the national project of imagining a better future for Latvians in Latvia, and a future in which Latvians are the primary shapers of their own destiny. For Rotbergs, the horse project and its possible follow-ups were not just about "restoring a European wilderness," but, what is more important, about resolving the very national problem of postproductivist rural development. Like Mayor Veits, he was deeply committed to keeping Latvians living and working in the countryside, albeit not primarily as farmers on dispersed home-steads. "Certainly the number of people employed in rural areas will shrink and so too must the amount of agricultural land," he told me. "But in the social sphere I would want to redirect all the money to rural children's education, to create op-portunities for them. I think this imbalance in Latvia between the countryside and Riga is completely unacceptable, this gap in development is terrible."[67] He of-fered various ideas for redressing the imbalance such as job training programs, tax incentives to promote extraurban investment, relocating government facilities in the provinces, and offering scholarships for rural students with a requirement to return home after graduation.

The crisis of rural development should have been vigorously debated nation-wide, Rotbergs lamented, but neither the state nor civil society had fostered this dialogue:

> Back in the Popular Front days [of the late 1980s], maybe some of these basic ques-tions were being raised about where we are going. But now they no longer are at all. . . . We have no intellectual leader. We should have had some respected fellow or some kind of process, or gotten together some kind of organization, a Club of Rome or what have you—that would be beautiful. The intelligentsia has completely [for-saken] its role . . . in tackling the basic questions of social structure—where can you read about that? I don't know, maybe in some obscure philosophical journal? It's just not there. There is not a single serious editorial column to be found.[68]

The Lake Pape projects, for Rotbergs, were important not only for their contribu-tion to biodiversity protection at a Eurasian scale, but also as a forum for stimulat-ing Latvians to grapple with the ultimate national question of "where we are go-ing." Nature restoration was valuable not only for its ecological benefits, but as a means for revalorizing the postproductivist landscape: for teaching Latvians to view abundant, "wild" land as a new kind of development asset. "You can't make some enormous public opinion campaign with the horses alone," he said, ac-knowledging that a single demonstration project was inadequate to achieve such a profound shift in consciousness. But he hoped that such projects could encour-age Latvians to think strategically about postproductivist rural development.

WWF's agrarian critics, however, did not share Rotbergs's faith in globalization of the ethnoscape as a national development strategy. Thus Jānis Priednieks of the Latvian Fund for Nature opposed the horse project as not only antiagrarian, but antinational:

> I believe that in the rural landscape it would be economically most efficient to subsidize agriculture, because first of all, the farmer will feed himself and his family, he will maintain the landscape, and something will be left over to sell. . . . Purely as a biologist, I am far from pleased that animals of fairly bizarre genetic origins are being imported entirely unnaturally into Latvia, and here you could even raise objections in terms of the Rio Convention [which prohibits introduction of exotic species]. . . . I would have voted for something of Latvian origins—like the "Blue cow" or the Latvian Brown, which also exhibit excellent qualities for surviving in winter conditions. . . . In the long-term perspective, I think it would have been better to subsidize the local farmers to keep the Latvian Brown. Because now an enormous territory has been taken out. Those animals, for all I know they could very well be dangerous. . . . That territory under [nature] management—maybe it's inexpensive, but at the same time people can't go in there. [The horses] can't be used for food, they don't produce milk, the meat will never be certified—it is of genetically unknown origins. So I'm far from pleased about this kind of management. But if a person can feed himself, keep a cow, mow, and herd, and if he gets subsidies for setting up a normal toilet, shower, and room for some tourists, I think that will cost the least and in the long run bring the greatest gain.[69]

Some of Priednieks's claims were factually inaccurate. As a descendant of the tarpan, which is known to have lived in the territory of Latvia, the Polish horse (Konik polski) is not truly an exotic species, and its genetic origins are well documented.[70] Far from being excluded from the project territory, people have been encouraged to visit (though preferably with a guide). As these fairly significant errors of fact suggest, he was speaking not "purely as a biologist" but also as an agrarian nationalist. From the agrarian perspective, sound nature stewardship could be rooted only in the age-old experience of the traditional Latvian farmer. To expel the farmer from his land was not only ecologically suspect but fundamentally antinational, for it signified that Lake Pape had essentially ceased to be part of Latvia, "an enormous territory has been taken out."[71]

Many local Rucavites, confronted daily with the grim realities of poverty and subsistence farming, shared WWF's pessimism (rather than Priednieks's optimism) about the economic future of farming. The librarian Gunta Timbra, despite her visions of development through defense of the ethnoscape, was largely resigned to the prospect of depopulation: "I have no illusions about the future for my children in our township. . . . We have only one way to go—completely to the city." The biologist Ināra Rūce held out the slim hope that the new global interest in nature would be able to fill the economic void: "We will have to be able to earn from nature, because very few will be able to earn from agriculture." But she, too, believed the "purest Latvians" were country folk: "Ultimately even city dwellers say—we come from the countryside. The best roots and the best thoughts come [from the countryside] . . . ; the farmstead [viensēta] is the Latvian lifestyle, mentality. But we are forced [to abandon it]." Rūce feared that by abandoning their ethnoscape, Latvians would eventually lose their very Latvianness. "I drill it into my children as much as I can that they are Latvians. But I think that sooner or later it will come to pass not that we will assimilate the Russians and other nations, but they will assimilate us. . . . It's a shame, but I don't know if we will still exist in a hundred years—Latvians who know they are Latvians."

The story of Lake Pape thus vividly captures the deeply rooted tension in Latvian national discourses between the outward-looking embrace of hybridity and the inward-looking embrace of landed labor. While the region has historically been an isolated "pickle jar" of local ethnographic uniqueness, sparsely inhabited by solitary farmers and fishermen, it has also long been permeated by international currents. As the Rucava Nature Fund pointed out in its almanac: "In the time of the Duchy of Courland [1562–1795], Rucava was no isolated corner, for the highway from Jelgava [capital of the duchy] to Klaipeda [in Lithuania] passed through it."[72] In the seventeenth and eighteenth centuries, Dutch weavers operated textile mills in Rucava, exporting the finished goods to Lithuania and Poland. And in the nineteenth century, plans were afoot to build a rail line from Rucava to Berlin. Indeed, Rucava's very geography invites a confrontation with the problem of whether to develop "by land or by sea," as suggested by the authors of the RNF almanac:

> The salty sea wind lashes the traveler when he stops, his back pressed against the Lithuanian border post at the sea's edge, to gaze into the distance. Here Latvia opens up like a great, never fully readable book, full of ancient writings and wonderful natural vistas. Rucava is a place where the sea meets the land—a gateway to the sea, a gateway to the land and to Latvia. But perhaps here begins another gateway through which we must cross—a gateway toward an awareness of the world, of nature, and of life.[73]

The ICZM process in the mid-1990s sparked a dramatic transformation in Rucavites' "awareness of the world, of nature, and of life." Though initiated by the forces of globalization in the form of the multilateral Liepāja Environmental Project, the developmental debate in Rucava was thoroughly intranational, structured by the parameters of the historic debate over Latvian identity.[74] Latvians articulated competing visions of both the endpoint to which the natural landscape should be "restored," and the "local characteristics and strengths" that should propel sustainable development. Those supporting the local cultural heritage, like many professionals at the regional and national level, looked to the land, to the "ancient writings" inscribed in the agrarian ethnoscape. Passionately committed to restoring a peopled landscape shaped by traditional farming and cattle herding, they hoped that somehow, with help from environmental subsidies for agriculture and heritage tourism, the ethnographic value of Rucava's landscape and the hard agrarian labor of its residents might revive the local economy. WWF-Latvia looked "seaward" to a new global narrative and sought to restore a pre-agricultural landscape by undoing the generations-old link between ecology and traditional farming—or, as the agrarians would have it, the synergy that linked ecology, farming, and Latvianness. Much as the nineteenth-century National Awakeners strove to emancipate Latvia from feudal bondage by reclaiming an ancient seafaring legacy and by tapping into the burgeoning markets and railways of the vast Russian Empire, so WWF hoped to reanimate the post-Soviet countryside by restoring preagricultural natural processes and engaging global markets for biodiversity and "European wilderness."

Those who have studied the aid process in post-Communist countries observe that the agendas of East European NGOs often reflect tactical responses to international funding opportunities.[75] But in the Lake Pape case, Latvian strategies were shaped as much by the competition between historically rooted local discourses. The Dutch strategy for nature development was not imposed on WWF-Latvia by donors. Quite the contrary: Rotbergs found the Dutch team himself and convinced his recalcitrant Swedish donors to hire them, thereby putting himself at odds with most of his colleagues in other West European branches of WWF, which remained largely wedded to the traditional agrarian approach.

Both WWF-Latvia and the Latvian Nature Fund compete for funds from international donors, and the pursuit of funding alone cannot explain their radically different agendas. Nor can it explain the zeal with which Priednieks of the LDF excoriated the horse project, nor the WWF's promotion of postproductivist alternatives to the agrarian conservation and development agenda in other efforts, sometimes in the face of tremendous hostility from peers. In other words, the Lake Pape case supports Brian Slocock's claim that "even though many western policy advisors entered central Europe firmly mounted on their white hobby-horses, . . . at the end of the day there is sufficient disagreement among western specialists to allow CEE policy-makers to 'pick-and-mix' from the advice offered and to shape the process according to their own interests and perceptions."[76]

At heart, the Lake Pape developmental debate was about the preservation of Latvianness in a context of increasing openness to the outside world. "The forests are here, but the money is there," as Rotbergs liked to say regarding Western donors. With investment capacity and market demand—for agricultural goods and nature amenities alike—so vastly greater in the West, the ultimate question posed by any rural development strategy was, as Mayor Dadzis of Jūrkalne put it: "Who will be in charge—we ourselves or others?"[77] Rotbergs hoped that with investments in building local capacity, Latvians would be able to manage the EU accession process to their own benefit, using Western interest in preserving Latvian nature both to protect continental biodiversity and to generate wealth. Much as Krišjānis Valdemārs was convinced that Latvians would remain Latvian even while aggressively pursuing the role of cosmopolitan middlemen, so Rotbergs believed that Latvians could globalize the ethnoscape without eroding their sovereignty or cultural distinctiveness.

In contrast to northern Kurzeme, where the proposed national park would have impinged on currently inhabited fishing villages, at Lake Pape WWF was creating the globally defined "new wilderness" in a space that had already been emptied under Soviet rule. The fishing villages were already deserted, and the land surrounding Lake Pape was already devalued "objectively and subjectively," to recall a favorite Soviet phrase; not only was it poor farmland, but no one was trying to farm it or use it for other extractive purposes. The horse project could be launched with the simple leasing of a few hectares from a single landowner, and while many residents viewed the project with skepticism and puzzlement, there was no outright hostility or opposition, for no one had anything directly to lose from it. At least for some residents, it was possible to imagine that Latvians could continue to preserve and display the agrarian ethnoscape while at the same time making room for "European wilderness."

At the time of this writing, the Konik horses have survived their seventh winter at Lake Pape, adapting successfully to a harsher climate and more predators than in the Netherlands. As was hoped, they have required little human assistance and drawn little overt opposition from local residents. Auspiciously for WWF, the herds' first foal was born on November 18, 1999—Latvian Independence Day. When I returned to Rucava in June 2002, the number of horses had grown to forty-four and appeared to be fulfilling their ecological function of enhancing biodiversity in the wet meadows through natural grazing. Some 2,000 tourists had visited thus far that summer, including several international groups (journalists from the Czech Republic, Austria, and even Canada; plenty of German motorcyclists; one Dane on a moped). Since my last visit, at least one more local resident had made the postproductivist leap: the full-time overseer and guide for the horse territory, hired in 1999 by WWF, had spent the previous seven years farming on her parents' land and had worked in Soviet times as an economist on the collective farm.

WWF-Latvia has continued to promote Lake Pape as wilderness. In September 2002, it opened an information center and guest house and a new seaside nature trail. In 2004, the last missing pieces of the Dutch planners' pyramid of grazers were introduced at Pape: wild cattle in February and European bison in June. Again, Rucavites did not resist these efforts, but WWF staff continued to lament their inability to persuade residents to take a more active role in developing the site's income-generating potential.[78] Local defenders of the ethnoscape, for their part, had not yet found a way to maintain working farms and make economic development "spring from cultural history." It remains to be seen whether either developmental vision can succeed in reviving the moribund post-Soviet countryside, and with what consequences for nature and people. In the aftermath of EU accession, as small-scale planning and demonstration projects increasingly give way to major subsidy programs and infrastructure development, how much autonomy will Latvians retain to contest the meanings of sustainable rural development on their own terms?

Chapter 7

"Lichens Are Not Our National Treasure"
The Battle over Sustainable Forestry

The central figure of Latvian agrarian identity discourse, as we have seen, is the small family farmer, solitary *saimnieks* of the dispersed farmstead, the *viensēta*. The iconography of the nation of farmers embraces not only literal tillers of the soil or herders of livestock, however, but also the rugged fisherman of the coastal hamlets, or any other Latvian who produces material value through an intimate encounter with nature. After the farmer, second place in the pantheon of agrarian nationalism must surely go to the state forester of the 1920s and 1930s. In his military-style uniform, living in a humble cabin deep in the woods, imbibing a profound and sensitive knowledge of the forest through daily traverses of his 500-hectare district, defending the new state's sylvan riches against desperate homesteaders and greedy capitalists, the forest ranger is a profoundly resonant symbol of the First Republic.

In the 1990s, a campaign led by WWF-Latvia to introduce contemporary Western sustainable forestry practices unleashed a firestorm of controversy over the proper stewardship of Latvia's "green gold." Drawing on recent insights from conservation biology and supported by allies recruited from the logging industry, WWF denounced long-established forestry practices as both ecologically harmful and economically irrational. The forestry establishment resisted this campaign bitterly, defending traditional stewardship principles in the familiar language of agrarian nationalism. Their opposition merged a productivist construction of nature with a nostalgic developmental vision hearkening back to the xenophobic statism of the golden age between the world wars. Against WWF's program of managing for natural forest conditions, traditionalists championed the orderly, parklike forest of the Ulmanis days. Castigating the newfangled "Swedish methods" advocated by WWF and its allies, they celebrated Latvia's "centuries-old" forestry traditions: traditions rooted in the German legacy of scientific forestry as well as in the Soviet legacy of utilitarian productivism and statism.

As Latvia ventured into its second era of independence, the forestry establishment still firmly embraced the silvicultural principles inherited from nineteenth-century Germany. These principles have long been upheld not only in Latvia but around the world, wherever the German school left its mark. In the United States, for example, as Nancy Langston demonstrates, in Oregon's Blue Mountains federal foresters attempted to "manage, perfect, and simplify the forests: to transform what one forester called in 1915 'the general riot of natural forest' . . . into a regulated, productive, sustained-yield forest." Their aim was not only to save the forest from abuse by industrial loggers seeking short-term profits, but also to improve "a 'natural' landscape that also needed saving because it was decadent, wasteful, and inefficient." They had absolute confidence in their own abilities to achieve this noble mission through the new discipline of scientific forestry. "As scientists who had the interests of America and American forests at heart, they felt they were beyond criticism."[1]

For several generations of American foresters, their fundamental productivist premise—that "the role of the forest was to grow trees as fast as it could"—made it hard for them to conceive of natural disturbances such as insects, disease, fire, and decay as anything but deadly threats to their treasured forests.[2] In the late 1980s, however, some American scientists began to argue for the positive ecological impact of such disturbances and to criticize the effects of traditional management. Champions of the New Forestry called for shifting priorities from maximizing yield to protecting biodiversity. By then, it was already too late to save the Blue Mountains from the unforeseen consequences of traditional management: replacement of magnificent old-growth ponderosa pine forests with thickets of fir and lodgepole that had been devastated by insect infestations and uncontrollable fires. But a debate had been opened that would help launch a global reconsideration of forest management. In the mid-1990s, WWF brought this debate to Latvian forestry circles.

Like the battle between advocates of biodiversity and traditional foresters in Oregon's Blue Mountains, the Latvian case was a struggle not between those who loved the forest and its enemies, but between two groups of forestry professionals, both of whom passionately believed themselves to be the forest's truest defenders. On each side of the battle line, claims about good forest stewardship were shaped by discourses of identity and ideas about Latvians' proper relationship to nature. Traditional foresters saw the forest as a vital element of the agrarian ethnoscape: orderly, cultivated, a living testimonial to the nation's history and the hard labor of the forefathers. Latvians, drawing on their rich local traditions, could plot the best future for themselves and their forests. WWF and its allies, for their part, could imagine no bright future for trees or people except through participation in global currents of commerce and science.

Defending the "Normal Forest"

During the Enlightenment, Prussian scientists developed a concept of forestry "largely based on controlling nature through mathematics and design."[3] Despite the concurrent stirrings of *völkisch* Romanticism, the "founding fathers of German forestry . . . did not at all like the idea that the forest is a gift of nature; on the contrary, they wanted the forest to become an artificial work of forestry."[4] Forests came to be viewed as objects of scientific manipulation, and central to the new scientific model was the notion of the "normal forest." Phil McManus writes:

> This was a mathematical model of trees of various ages growing and being cut when they reach the "mature" stage, and the area being replanted to enable growth for future cutting. . . . Nature was no longer seen as wilderness, or unmanaged trees of various ages growing together. Instead, blocks of even aged trees, growing on shorter life spans than old-growth forests, became defined as "normal."[5]

As Aldo Leopold put it, the normal forest reflected German foresters' "passion for unnecessary outdoor geometry." "Most German forests," Leopold wryly noted after traveling there in 1935, "would do credit to any cubist. The trees are not only in rows and all of a kind, but often the various age-blocks are parallelograms. . . . The boundary between wood and field tends to be sharp, straight, and absolute, unbroken by those charming little indecisions . . . [that] bind wood and field into an harmonious whole."[6]

German scientific forestry made its way to the Baltic provinces in the late 1700s, when great landowners sought to emulate the latest German fashion by carrying out forest inventories and implementing the geometric system of felling trees by age blocks.[7] Foresters trained at German and Russian universities brought forest science to the Baltikum with the founding of the Baltic and Kurland foresters' societies in the late nineteenth century. The first forest inventory bureau in Latvian territory was established in 1907, and a classification system for Latvian forests was developed in the 1910s. Throughout the nineteenth century, however, growing numbers of ethnic Latvians who had been trained as foresters had to seek employment outside the Baltic provinces, thanks to the discriminatory policies of the Baltic German ruling class. Not until the early twentieth century did the first ethnic Latvian forest rangers begin to serve in Latvia.

In the agrarian reform of the 1920s, the government of newly independent Latvia expropriated forest estates and either parceled them out to homesteaders or transferred them to the state forest fund. An administrative system of forestry districts was established, with three levels of civil servants—rangers, foresters, and chief foresters—charged with managing state forests and overseeing private forest use. A Latvian forestry students' fraternity was founded in 1923 and the Latvian Forest Workers' Association in 1924. Foresters of the interwar period largely remained faithful to the norms and practices of German scientific forestry, but infused them with Latvian nationalist significance. State foresters and politicians identified beautification of the landscape with patriotism in the public ritual of Forest Days. Given Latvia's hard-won and precarious statehood, foresters regarded forests as the new state's "only natural treasure" and cast themselves as the sole defenders of Latvia's "green gold" against misuse and overharvesting.

In their new journal, *Forest Life*, state foresters recounted epic struggles to defend Latvia's forests from exploitation by well-connected politicians as well as hundreds of thousands of homesteaders seeking to rebuild war-ravaged farms or establish new ones. Farmers were issued lumber from state forests at subsidized prices; as a result, according to Heinrihs Strods, "Latvia's forests 'carried' . . . the whole weight of Latvia's agrarian reform, rural construction, and the renewal of the Latvian cultural environment."[8] The inaugural issue of *Forest Life* opened with a salvo against overexploitation:

> Our forests have many enemies. In this postrevolutionary era, when the so-called fruits of the revolution are being divided up, many have come to covet our forests, as the greatest natural treasure gained by our nation. . . . Insofar as the parceling out and use of forests is dictated by life's necessities, forest workers will support it. But when it becomes a form of unjustifiable wastefulness, when particular political parties, making unfulfillable promises at the state's expense, seek to benefit only their own leaders, and when annihilation of the forest begins to threaten future generations' economic foundations, then forest workers must cry "hands off" the nation's property.[9]

"We have been appointed as the guardians of Latvia's forests," declared a participant at the second Latvian Congress of State Foresters in 1922, "and we will guard them, and if the moment comes . . . when the last Latvian pine is felled, then may God protect us if we are responsible."[10]

State foresters described themselves as uniquely capable of stewarding this national patrimony, given their specialized training and knowledge and their ability to take the long-range view necessitated by forests' long growth cycle. "The forester works for the future and rarely sees the results of his labor," observed one commentator, noting that this difficult work was made even harder because the rest of society viewed the forest only in terms of private short-term gain. The chief weapon of professional foresters was their intellectual wealth, agreed another commentator, noting: "There is no other class of professionals like us, who are entrusted with such great state wealth and who must labor solely among country folk—the least educated part of the population."[11] The special competence of state foresters derived also from their closeness to nature, from living in the heart of the forest, far from other people, where "a person grows together with nature, understands it, and fully senses its power. He grows accustomed to the quiet life in the forest, enters into the forest's soughing and its harmony, understands its soul."[12] Evidence of this perennially close relationship was adduced from depictions of forest rangers in the traditional folk songs.

After the Soviet annexation, privately owned forest lands were nationalized and state forest management—from tree planting through felling—was taken over by integrated forest industry enterprises (mežrūpniecības saimniecības) under the purview of the ministry of forestry and forest industry.[13] The state owned roughly two-thirds of forest lands and managed and logged them through these enterprises; collective and state farms owned most of the remainder. War, emigration, and deportation had decimated the top ranks of forestry professionals, but the survivors remained faithful to the principles of German forestry. The Latvian SSR Forest Inventory Enterprise planned all final felling in narrow 100- or 50-meter clear-cuts, in strict accordance with felling-age theory and with the geometric forest inventory grid. Planners took the German predilection for uniform stands to new extremes with the promotion of high-yield monocultures beginning in the so-called spruce era of the 1950s and 1960s. Mixed stands of trees were discouraged on the principle that "the presence of birch, aspen, and white alder in a coniferous forest is not an indication of a natural, stable forest, but rather attests to a natural catastrophe or improper forest management."[14] Drainage of wet forests was greatly intensified to increase timber productivity.

After independence in 1991, the new State Forest Service renewed the interwar system of forestry districts. Along with other institutions banned by the Soviets, the Forest Workers' Association resurrected itself and its journal, *Forest Life*. In an early attempt to break up the state's monopolistic control, the government transferred the logging portion of the forest cycle to the private sector, disbanding or privatizing the Soviet-era enterprises. Forestry districts began leasing and auctioning short- and medium-term logging rights in state forests to private firms. Silvicultural management remained in the hands of the forestry districts, however, and even private logging firms were required to follow the felling plans issued by the State Forest Inventory Institute. Added to the responsibilities of state officials was oversight of private forest use, as the third wave of agrarian reform launched the complicated process of restoring private parcels to their former owners or their heirs. Restitution was still ongoing at the end of the 1990s, and officials estimated that by the time the process was completed, up to half of Latvian forest lands would be in the hands of 150,000 to 200,000 private owners.

Forests have played an important role in Latvia's economic recovery since independence, with forest products consistently ranking as Latvia's leading export group, growing from 24.4 percent of total exports in 1996 to 33.5 percent in 1998.[15] "Latvia's oil is timber and forests!" trumpeted a characteristic headline in Diena in 1997.[16] At the same time, forests—and especially the more than 20 percent of forests that are on wet soils—are among Latvia's richest habitat resources. Given this confluence of economic and biodiversity value, it is not surprising that WWF-Latvia targeted its most aggressive campaign for sustainable development and protection of biodiversity at the forestry sector. This campaign unleashed a fierce debate among Latvian scientists and foresters over the proper stewardship of Latvia's green gold.

The WWF Campaign

On the face of it, biodiversity in Latvia's forests had not been unduly threatened in the years following independence. Because so many farms were abandoned as a result of Soviet collectivization and urbanization, Latvia's total forest cover had grown by more than a million hectares—from 25 percent to nearly 45 percent of the country's territory—since the interwar period, and this trend was continuing, thanks to agricultural collapse and to the annual cutting limits enacted by the parliament. Twenty-five percent of forests enjoyed a range of protected designations that restricted logging. By many accounts, Latvian forests were adequately protected from overharvesting and loss of biodiversity.

Nevertheless, in 1992 Uģis Rotbergs, project manager for WWF's newly founded Latvia program, published an editorial warning of the destructive impact on biodiversity of modern industrial forestry practices, widely employed in Latvia, such as drainage and monoculture plantings. Citing the Rio convention on biodiversity, recently signed by Latvia, Rotbergs called for a fundamental reorienting of Latvia forest policy to moderate the focus on maximizing output with a new commitment to conserving biodiversity. "In shaping future forest policy in Latvia," he declared, "to be guided solely by the price of timber in Europe and by forest productivity (that is, timber productivity) is not only short-sighted, but also dangerous."[17] Instead, Rotbergs urged Latvian foresters to embrace Western principles aimed at balancing the economic, social, and ecological functions of forests.

Ecology: From the Normal Forest to Protecting Biodiversity

On the ecological front, according to WWF, the root of the problem was that Latvian foresters, during the enforced isolation of the Soviet period, had remained wedded to the principles of the "normal forest," while Western forest management had been radically challenged by the "new paradigm" in conservation biology. In Western Europe and North America, ecologists have sought, with varying degrees of success, to mitigate the productivism or resourcism of traditional forestry with a new focus on wildlife preservation. Embracing the notion of nature as flux and dynamism, the New Forestry maintains that the best way to preserve wildlife habitat is to encourage natural processes to the fullest extent.

The first step toward this goal is to make a careful inventory not simply of tree species and growth conditions, but of the full panoply of forest species, habitats, and ecosystems. In this view, German-style scientific forestry, with its clear-cuts, same-age tree stands, monoculture, drainage, firefighting, and so on, inhibits natural processes and reduces natural complexity, impairing the ecosystem. Especially vulnerable are the habitat specialists that require specific conditions to survive, such as eagles, which nest only in very tall trees, or lichens that grow on rotting logs. Many conservation biologists argue that the loss of these habitats cannot be adequately compensated simply by setting aside protected nature areas; rather, "site-adapted" methods must be incorporated into the management of commercial forests as well. Old, decaying, and fallen trees should be left in clearcuts; logging remains should be chopped up, rather than hauled out or burned; cuts should follow natural landscape or ecosystem boundaries; regeneration should occur naturally or should at least promote mixed stands rather than monocultures. Forest drainage should be avoided because of the great value of wetlands as habitat, while practices that mimic natural disturbances, such as controlled burns, should be encouraged. The goal of forest science and management, in short, is to identify "natural forest" conditions and to approximate them as closely as possible.[18]

In the early 1990s, however, forest management in Latvia had not even begun to grapple with the new paradigm. As Rotbergs and other conservation professionals pointed out, there was no systematic ecological data on Latvian forests, since the inventories conducted from the 1700s through the 1990s had been essentially timber inventories, cataloguing only tree species, size, age, class, and so forth. Since the late 1970s, data on species and habitats had occasionally been included in forest inventories, but only haphazardly and at the initiative of individual biologists.[19] In the absence of systematic data, the designation of protected forest areas had little grounding in ecology, and thus there was no guarantee that the most ecologically significant sites were protected.[20] The *National Biodiversity Action Plan for Latvia* prepared in 1995, for example, reported that only 27 percent of black stork nesting sites were in protected areas.[21] Moreover, protected status did not always bring ecologically significant logging restrictions; in most cases the protection regime involved nothing more than delaying final tree felling by one age class, and logging was banned entirely in only around 1 percent of Latvia's forests.[22]

In short, the fact that 25 percent of Latvian forests were included in some type of protected category offered no assurance of actual defense of biodiversity. According to Rotbergs and other critics, Latvia's high levels of biological diversity relative to the West were not evidence of effective conservation policies, but rather the accidental by-product of fortunate hydrology (the preponderance of wetlands) combined with late development and Soviet inefficiency. As veterans of the Soviet period noted, overharvesting was prevented not only by formal logging restrictions, but also because the planned economy encouraged people to reduce their workloads by negotiating for lower output quotas.[23] In this sense, as one official wryly put it, "socialism was very good to fauna."[24]

Genuine biodiversity protection, according to WWF officials, required not only rationalizing the protected areas system, but also implementing site-adapted management practices in commercial forests. The problem, however, was that Latvian forest management regulations not only discouraged such practices, but in many cases expressly forbade them. The Forest Inventory Institute would not sanction deviations from its "cubistic" logging plans in order to follow natural landscape boundaries, for example, and state foresters imposed fines for leaving uncut trees or logging remains in clear-cuts. Moreover, state-issued management plans were targeted at the individual stand, rather than the ecosystem, thereby promoting fragmentation of natural corridors and threatening the integrity of the ecosystem.

The Mežole Project

The centerpiece of WWF's campaign to integrate biodiversity protection into Latvian forest management was a demonstration project at Mežole, in the Smiltene district of northeastern Latvia. Funded by the German Federal Ministry for Economic Cooperation and Development (BMZ), the Mežole Project was launched in 1995 and concluded in December 1998. In the summer and fall of 1997, all forest stands in the 15,000-hectare project territory were inventoried using a forest structures checklist methodology adapted from Swedish practice. Local residents and forestry district staff were trained in this methodology, as well as in the use of Geographical Information Systems (GIS) technology. Ecological data were combined with existing socioeconomic and timber data in a GIS data base, from which an ecological landscape plan was developed for use in forest management planning. Wetland systems and riverine dispersal corridors were identified and mapped, and important habitats were linked in local networks. Several demonstration sites were selected for site-adapted methods, such as controlled burning, new to Latvia.[25] The ecological landscape plan was submitted to public review at meetings with local authorities, logging firms, and area residents and was distributed to various institutions and scientists nationwide.

WWF sponsored research by the biologist Normunds Priedītis to identify areas of high biodiversity in wet forests throughout Latvia and to make recommendations for their protection. Priedītis's research was incorporated into a forest ecology handbook for forestry officials, *Latvian Forests: Nature and Diversity*, produced by a desktop publishing facility established at Mežole.[26] As WWF-International representative Magnus Sylven noted, it was "the first book of its kind in Latvian addressing people from the forestry sector, forest biologists, students, and any other persons interested in forest biota and sustainable forestry."[27]

With funding from WWF-Denmark, a team of Latvian biologists employed the Mežole ecological inventory methodology to analyze forest structures in four sample areas representing diverse biogeographical conditions. This project culminated in a scientific monograph, *Natural Forests of Latvia*.[28] Other materials published at Mežole included a paper on the political challenges to sustainability-oriented forestry reform, which circulated widely in forestry circles;[29] a forest ecology teaching guide, which was distributed to every public school in Latvia; and a series of pamphlets on forest ecology, each describing a particular habitat type, its developmental dynamics, and policies for biodiversity management. Seeking to redress the inadequacies of the Latvian forest classification system, WWF distributed the pamphlets to all forestry district offices and forestry schools, some of which incorporated them into their curricula.

Over the course of the project, at least 1,300 people attended field tours or seminars and training courses in forest ecology, GIS technology, and other aspects of sustainable forestry. State officials were the project's primary audience, but the project also attracted logging firm executives, scientists, government staffers, and international visitors. This was Latvia's first forum for focused debate between advocates and critics of the Western sustainable forestry paradigm. During its four-year course, the Mežole Project became, according to external evaluators, "clearly the major talking point in relation to forestry and forest management in

Latvia."[30] And for those most directly involved—project staff, local forestry district officials, and logging executives—the Mežole Project represented something of a conversion experience, as their enthusiastic testimonials suggest. One reported, "The project was like throwing a stone in the pond—it started the discussion. Before then, unshakeable axioms governed everything. The project opened, publicized the discussion."[31] The chief engineer at the Silva logging firm testified, "The Mežole Project gave us knowledge and new instruments. . . . It gave us the opportunity to learn how to evaluate the forest in terms of the three criteria: environmentally friendly, economically viable and socially equitable. The project was the first to begin talking in this way."[32] Said a project assistant, "All the contacts with specialists from abroad, from both East and West . . . created many opportunities. The work made you think harder and look for the roots of problems."[33] And a state forester reported, "The amount of information that came through the project was enormous. Before the project we looked at the forest in a completely different way."[34]

The first two statements quoted above were made by logging executives, and indeed logging firms proved to be among the most enthusiastic converts to this new way of "looking at the forest." According to an external midterm review, firms operating in the project area reported savings of 20 to 30 percent as a result of practices banned elsewhere in Latvia, such as natural regeneration, leaving dead wood and brush in logged areas, avoiding boggy areas, and, thanks to the landscape-level planning approach, building fewer temporary logging roads.[35] More broadly, two factors aided WWF in convincing Latvian loggers of the virtues of sustainability: the perverse regulatory environment in Latvia and the growing global clout of the green consumer.

Economics: "The Civil Servant Is God"

The root of the problem on the economic front, WWF and loggers concurred, was the onerous Soviet legacy of excessive state intervention in commercial forestry. Firms operating in state-owned forests found themselves hampered in several respects. First, only short- and medium-term logging leases and auctioned permits were available. In the absence of extended contracts, firms lacked the security to make long-term investments or engage in long-range planning, and their logging sites were typically dispersed and fragmented. Second, while logging had been privatized after independence, silvicultural management—regeneration, thinning, and so on—remained in the hands of the state forestry districts, so that control over the forestry cycle was fragmented. Finally, the state directly micromanaged firms' logging activities not only through the detailed management plans drafted by the Forest Inventory Institute, but also through countless supplementary "management directives" issued by forestry districts. Management plans and directives mandated not only how much a firm had to cut from its total allotment over the term of its contract, but specifically what to cut each year in each forest compartment.

In short, the state retained nearly all of its Soviet-era monopoly control over forestry. Forest Service officials drafted and implemented regulations; the Forest Inventory Institute issued mandatory management plants; state foresters and rangers directly carried out silvicultural activities and micromanaged logging by private contractors. All this necessitated a very large apparatus of civil servants: around 200 central agency employees and over 2,300 state foresters and rangers in a country with 3 million hectares of forest. By comparison, Rotbergs liked to point out, Sweden managed its 24 million hectares of forest with only 1,100 civil servants.

In the Soviet period, this army of civil servants had been funded directly from the state budget. After a 1994 reform, however, the forest sector received the bulk of its funding directly from the sale of forest resources and products, and primarily from the sale of logging rights to private contractors. By many accounts, this system encouraged extortion by state foresters. Through so-called racket contracts appended to their leases, for example, firms could be forced to pay forestry districts for fictional services. According to many logging executives, corruption and incompetence among state foresters posed the gravest obstacle to good forest stewardship. As one observer put it, "Today the civil servant is God, even more than in the Soviet days."[36] The budgetary system impeded firms' operations more fundamentally, as well. The geometry of the normal forest dictated the cutting of every tree in a compartment, and chief foresters now had a strong material incentive to enforce this policy strictly, for their own field offices were funded directly from stumpage fees paid by loggers per felled tree, as well as from fines collected for leaving trees in clear-cuts and other violations of management directives.[37] Thus, according to Gunārs Dišlers, chief forester in the Ogre district, the chief source of conflict between forestry districts and loggers was the loggers' failure to harvest their entire contracted allotment. In 1994, for example, Dišlers broke his contract with Silva, a Latvian-Finnish joint venture and the largest commercial logger operating in Latvia, because in the previous year Silva had harvested "only the best stands" but left over half of the total allotment uncut.[38]

WWF and its logging allies argued that this system was ecologically destructive because it forbade the use of the new site-adapted methods. But it was also economically inefficient because it prevented loggers from making market-based decisions about when, where, and how much to log—concentrating their logging in one area to minimize transportation costs, for example, or carrying out commercial thinning only when pulpwood prices were high. The irrationality of the system was made manifest in the very terminology of Latvian forest management planning: because all trees in a given compartment had to be cut to preserve the rectilinear geometry of the normal forest, some trees had to be cut before they reached final felling age, and these were officially designated as an "economic sacrifice." In other cases, the directives compelled a firm to do environmental damage, if, for example, in a last-ditch effort to meet its quota by the October deadline, it had to cut during the rainy season rather than waiting for the winter freeze. According to one firm's chief engineer, Arnis Melnis, these enforced irrationalities directly affected profit margins, contributing to layoffs and volume reductions in 1996.[39]

The system threatened to generate even greater economic losses in the future, thanks to the growing consumer demand for environmentally friendly wood products, particularly in Germany, the United Kingdom, and other West European countries. Spurred by market signals, in 1993 major logging firms had joined WWF-International and other environmental organizations in an unusual coalition to devise an international system of green forest certification overseen by the Forest Stewardship Council (FSC), an independent body. Nordic and particularly Swedish firms had led the way in implementing site-adapted practices in their own forests, but when they attempted to do so in their joint ventures in Latvia, they incurred fines and risked losing their contracts. Moreover, Nordic wood processing firms were major importers of timber from Latvia, as well as from neighboring Estonia, Lithuania, and Russia. Without certification for their eastern inputs, these firms feared that eventually they would become uncompetitive in German or British markets.

Latvian loggers, in short, had ample economic incentive to embrace WWF's vision of sustainable forestry, and thus by 1997 the Mežole Project's external reviewers could assert: "Forest companies are now clearly major allies in relation to the broader aims and objectives of the project."[40] In 1997 Silva, Latsin, and three other large private logging companies, which together comprised 20 percent of Latvian logging capacity, joined WWF-Latvia in establishing a Forest Club to lobby for sustainability-oriented policy reforms. Following the recommendations of a Swedish-supported forest sector review conducted in 1995, the Forest Club called for ending direct state intervention in forest management and logging and for limiting the state's role to legislation, enforcement, and facilitating private sector development.[41] Logging firms, they argued, should be entrusted with long-term leases and with responsibility for the full forestry cycle. In their view, freeing firms to respond to global market signals and to participate in the global green certification system was the most effective means of ensuring an economically vibrant and ecologically responsible forest sector in Latvia. The number of rangers should be greatly reduced: technologically well equipped and freed from the burden of micromanaging private logging activities, a smaller number of rangers would be able to monitor much larger districts. Given a favorable climate for private enterprise, state employees laid off through such downsizing could find jobs with private forestry firms.

WWF also called for a change in the state's relationship to Latvia's private forest owners. After a fifty-year breach in landownership, most new owners lacked experience and knowledge of forest management, and because of poverty, many were exploiting their forest holdings heavily to make ends meet and keep their farms afloat. Many forestry officials saw private owners as unable or unwilling to manage their forests responsibly and believed that only tough state sanctions could prevent depletion of forest resources. This attitude, together with certain lacunae and inconsistencies in the legal definition of property rights and obligations, led some private owners to fear growing state harassment or even renationalization of forest properties. WWF called upon state officials to tackle the challenge of private forest ownership not by intensifying sanctions and penalties for misuse of forest resources, but rather by supporting education and outreach for new owners. According to WWF, sustainable forestry could best be promoted by fostering an economically secure and environmentally aware class of small private owners.

In envisioning the future development of the Latvian forest sector, in short, WWF looked to the West and its valorization of biodiversity, private enterprise, and global markets. For Rotbergs, the future was Western:

> Right now we have a unique opportunity to look into the future. By observing our Western neighbors, we can say with reasonable certainty that they live at least fifty years ahead of us. Through intensive forestry development, they have transformed the majority of their forest ecosystems. Not only have plant and animal species disappeared, which are at present still common here, but so have entire habitats. Perhaps it is the longing for bygone times that drives foreigners to admire our nature. But we might also conclude that several decades from now we will be like them and we will want to hear the oriole's song or see the stork catching frogs in the meadow.[42]

Future Latvians, in other words, would see the forest the way Westerners now do.

In Rotbergs's Valdemārian developmental vision, moreover, Latvia's conversion to sustainable forestry could itself be a valuable asset to be traded in the global marketplace of development aid. As the Mežole Project's final report noted: "Mežole provides the only living model of sustainable forestry development in a post-Communist legal and economic environment," and as such it constituted a unique resource "for the promotion of sustainable forestry through the entire central and east European region." In the course of the WWF project, Mežole hosted 115 visitors from Western countries and 125 from the East, including Estonia, Lithuania, Belarus, Georgia, Russia, Ukraine, and Hungary. "It was only after visiting Mežole," the final report claimed, "that forest officers from the Russian Far East agreed to join the FSC certification process."[43] As Magnus Sylven of WWF-International put it: "Being the European country most dependent on the forest sector, Latvia could probably serve as a 'model country' for sustainable forestry development. The commonalities of the post-Soviet era, the applicability of the Russian language, and WWF policy work in Latvia makes the WWF forest team a potentially interesting resource."[44] Rotbergs hoped to attract international donor funds by marketing Latvia as a "low-hanging fruit" among the former Communist countries. Much like Krišjanis Valdemārs, in other words, he believed that Latvians could prosper best by serving as middlemen between the financial resources of the West and the natural resources of the East. As he liked to say, "The forests are here, but the money is there."

"We Like *Ordnung*": Traditional Foresters Fight Back

The WWF-led sustainability campaign was met with tremendous hostility by Latvia's mainstream forestry establishment, including a majority of forestry officials and academic forest scientists, as well as some loggers (typically those affiliated with recently privatized Soviet-era enterprises. Identifying maximum timber production as the sovereign goal of forest stewardship, these critics defined sustainability strictly in terms of productive resources. As Soviet-era forestry minister Leons Vītols put it: "Good forestry means managing for the maximum growth in a given forest area."[45] Pēteris Zālītis, a veteran forest scientist and outspoken critic of WWF, concurred: "Sustainable, in my opinion, means continuous, increasing forest production, such that I can log every year, and log more every year."[46] From this perspective, guaranteeing maximum yield was a fundamental obligation of the state. Zālītis noted that he had been campaigning since 1990 for "the necessity of declaring at the level of fundamental state principles the self-evident truth that land, including forest land, must produce output."[47]

Nature protection, on the other hand, was to be indulged in at the margins, in special areas set aside for that purpose. "When logging and managing our forests," Zālītis told me, "we have always protected special little places, little stands, tree clusters, or some kinds of landscapes." Defining the value of Latvia's forests entirely in terms of material production, traditionalists ridiculed the value placed on biodiversity by WWF. "Lichens are not our greatest national treasure," declared Zālītis. "We can only regret that right now activities are being supported with European and partially with our own country's funds, the essence of which is expressed in the widely propagated slogans: 'Latvia's degraded environment is our greatest treasure!' or 'Our goal and pride—as many rotting trees as possible in Latvia's forests!'"[48] Drainage of wet forests was perhaps the most critical bone of contention between the productivists and the advocates of biodiversity. The extensive forest drainage system from the Soviet era had been largely abandoned since independence owing to lack of funding, and whereas WWF was deeply committed to preserving Latvia's wet forests as a unique source of biodiversity, traditional foresters were equally committed to reviving drainage as the most important tool for increasing timber productivity. "We must stop talking nonsense about the great ecological value of wetlands," maintained Mārtiņš Dāboliņš, a veteran Forest Inventory Institute technician. "They are valuable in the Sahara, but here we're up to our necks in water in the fall, and the farmer can't even get out in his field."[49]

A corollary of this productivism was the understanding of forests not as wild nature but as cultivated landscapes of labor—a view inherited from German scientific forestry but also reinforced by Latvian agrarian nationalism and Soviet utilitarianism. Thus Zālītis maintained that present-day biodiversity levels in Latvian forests were the product of human cultivation: "All of this, these storks and so forth, that everyone is so excited about—it is all linked to human goal-oriented activity, and not simply God-given. . . . The forest is not a *natural* treasure: the forest is *national* treasure, it is cultivated." Therefore, he told me, "contemporary forestry must start from the position that the forest is an inextricable element of the human-inhabited and human-managed environment. The vision for the future cannot be based on untouchable nature, with man behind a barbed-wire fence, playing with paper flowers." For Zālītis, in other words, the WWF vision of site-adapted forestry was an attempt to exclude man from nature, and as such it was implicated in transforming the agrarian ethnoscape into the post-Soviet landscape of abandonment. Traditionalists denounced this perceived exile from the productive landscape of labor in classic agrarian nationalist terms. "Land is the foundation of every Latvian's life, and the forest—his savior in times of trouble," declared a forest scientist. "To take land away from a Latvian is the same as knocking the ground out from beneath his feet. Destroy the forest, and the Latvian nation will disappear, scatter without any external coercion."[50] Another wondered, "Must Latvians really live in nature reserves, among key habitats of value for animals, and subsist on humanitarian aid, while at the same time not exploiting for their own needs the country's chief treasure—the land and the forests cultivated on it by people?"[51]

Traditionalists assailed site-adapted forestry not only because it reduced timber production but also because it violated the aesthetic tradition of landscape cultivation. "Once upon a time, in the days of our fathers' fathers, one had to clear out the last little branch, almost with a broom, when cleaning up after a cut," mused chief forester Dišlers.[52] Leaving cutting remains and so-called ecological trees in clear-cuts was seen as violating the hallowed tradition of maintaining a parklike forest. The well-managed forest by traditional Latvian standards was indeed the very inverse of WWF's "natural forest." Hence the following morality tale recounted by Aija Zviedre, president of the Forest Employees' Association (and displayed ironically on the wall of Rotbergs's office at WWF):

> I recently witnessed a miracle. In the middle of a large forest tract, two pensioners had regained ownership of a 14-hectare parcel. Last winter they had driven 30 kilometers nearly every day to tend it. And not for income! They had gathered fallen, rotting wood, cleared undergrowth, and cut down dead trees. Everything worthless was burned, and firewood was neatly stacked, even though no one needed it. The several hectares of tended forest looked like a park. Other owners' properties were impassable jungle, but this forest had little berry-bush clearings, mushrooming paths, birch clusters, sunny corners and—joy. People like these need only recognition and a ranger's advice.[53]

From this perspective, the WWF model could only be lamented as poor stewardship. "I can't understand why negligent forestry should be justified with scientific pronouncements," complained former minister Vītols. "The kind of forestry where if I had been a forest ranger or what have you, I would have lost my job. The kind where you take only what you need and leave, say, a hollow aspen. . . . I understand forestry differently. I understand forestry as being where order is restored after every activity." When I asked Forest Inventory Institute director Jānis Vazdiķis why he opposed departing from the geometric grid to follow natural boundaries in planning cuts, he invoked this sense of Germanic orderliness, declaring wryly: "We like ordnung!"[54]

Indeed, traditionalists self-consciously linked the Latvian aesthetic of ordered nature to the German forestry legacy, habitually appealing to "centuries-old forestry traditions." "We have very deep roots, roots from the German barons here," explained Juris Matīss, director of the Forest Inventory Institute during the Soviet era.[55] For traditionalists, contemporary Latvian forest science was unimpeachable because it had remained faithful to the principles adapted from the German model a century ago. "We have the most brilliant system for classifying forest growth conditions—the world's best, as far as I know," asserted Matīss confidently.

Traditionalists maintained that this German-rooted system already contained an adequate understanding of forest ecology. "Truly, there is no other country in the world where the forests have been managed for so long in keeping with ecological prerequisites, ecological laws," insisted Zālītis. "These ecological 'rules of the game' are included in our system of forest classification." A Forest Service official agreed: "All of Latvian forestry is based on ecological principles. . . . We have our own school, our own methodology."[56] Despite WWF's assertions to the contrary, argued Vazdiķis, "we have really always operated with ecological correctness; we just need to know how to advertise ourselves." Traditionalists linked this "ecological correctness" not only to the German legacy but also to the primordial Latvian sensitivity to nature posited by agrarian nationalism. WWF might present itself as a pioneering advocate of biodiversity protection, but in Latvia, declared Zālītis, "our relationship to the forest was already formulated in the Latvian folk songs, and no one will ever formulate it any better. We don't need the Danes, the Swedes or the Finns—no one will outdo it."

The other side of this nostalgic celebration of Latvian forestry traditions was a xenophobic hostility to Western-imported notions of sustainable forestry, disparagingly referred to as the "Swedish methods" and derided as a slavish imitation of the latest West European fashion. Latvian foresters already practiced sustainable forestry, the traditionalists argued, but "our" sustainability was different from "theirs," because natural conditions were different in Latvia. "A Swede can't evaluate our forest better than we ourselves can," asserted the director of a privatized forestry enterprise.[57] "We must abandon our characteristic Latvian humility, free ourselves from our inferiority complex," urged Zālītis. "I can assert that we truly know enough about our forests, about their 'rules of the game,' to manage them rationally."[58] Perhaps sensing the irony of invoking a German legacy to denounce Western imports, Zālītis claimed that in comparison to the present situation, the earlier copying of German practices was less slavish: "Despite their German origins, Latvia's first forest scientists did not attempt to extrapolate their school learning mechanically to Latvia, but cleverly grasped the unique features of Latvian forests."[59]

For traditionalists, in other words, classic Latvian forestry and Western sustainable forestry were distinct but equally valid scientific models, and if anything, it was the Western approach that was scientifically suspect. Zālītis had even developed his own idiosyncratic counterinterpretation of forest ecology. Citing as the "basic premise" of ecology the cybernetic principle that "a system is better, the more energy it contains," he argued that managing for natural forest conditions "actually promotes ecological degradation" by reducing the flow of energy into the ecosystem by human labor—such as drainage, road building, and protection from pests. What Western forest ecology defined as natural, in other words, Zālītis denounced as degraded. Latvia's wet forests represented not a natural condition, in this view, but rather the result of an invasive process of "swampification" that "strangled" forests and reduced their ecological and economic value. To buttress this claim, Zālītis published studies in Latvian journals employing mathematical modeling to demonstrate that drainage increased biodiversity.[60] For Zālītis, Western management for "natural forest" conditions was a retreat to a more primitive past, a turning back of the clock on scientific progress: "If I am told that the forest must be burned, that the pine forests we see today in Latvia developed in connection with a degraded environment, and that we must further degrade this environment in order to preserve these pine forests—then some kind of abstract model of the past is being put forth as a management goal."

In addition to rejecting Western theories of forest ecology, traditionalists suggested that the "Swedish methods" were not grounded in ecological science at all. Instead, they were a cynical ploy on behalf of Latvia's Nordic competitors who wanted Latvia to lower its timber output in the name of ecology, and private logging companies, both foreign-owned and domestic, who hoped to benefit from reduced costs. According to the traditionalist skeptics, these savings would spring not from a happy coincidence of ecological and economic rationalization, but rather from the pursuit of short-term private gain at the expense of sound forest stewardship. As Matīss of the Forest Inventory Institute flatly asserted: "The goal of the Mežole Project is to make it easier for loggers to work. It's easier to leave those trees that they don't need." Zālītis was more explicit:

> Wealthy logging firms hire ecologists to advertise as ecologically necessary any activity that is convenient for the firm. Thus activities that up to now have been legally proscribed or socially shunned in Latvia gain a pseudo-ecological justification: littering clear-cuts, leaving rotting trunks and sickly tree groups, the exaggerated celebration of wet forests and natural regeneration, the creation of mixed pine-aspen stands, which is absurd from a forestry point of view. . . . Thus the depleters, the destroyers of Latvia's forests, find support in the World Wide Fund for Nature. . . . I cannot agree with these pseudoscientific formulations about the crucial role of Latvia's forest lichens in the life of the entire planet, which are being put forth on behalf of the immediate interests of logging firms.[61]

The State Forester: Steward of the Nation's Property

Underlying this profound skepticism toward private loggers and "Swedish methods" was the notion, inherited from the interwar era, of state foresters as a special brotherhood endowed with a unique sense of patriotic responsibility for Latvia and its forests. In the impassioned words of veteran forest scientist Miervaldis Bušs:

> There are few professions so firmly united by a responsibility for the near and distant future of the nation, for its material well-being and spiritual health, as are foresters. From this sense of responsibility comes a sense of unity, a capacity to think in the categories of the future, to rise above the mundane squabbles and base avarice that some try to exploit like an evil fuel to promote "progress" in certain other sectors. The forest "filters out" its people, teaches them the laws of eternity, forms its own worldview, and keeps hold of them. Not everyone is fit for the forest.[62]

Traditionalists located the roots of this unique esprit de corps in the interwar period, when Latvians, trained in the metropoles of Germany and Russia, laid the foundations of professional forestry and won high social esteem in the newly independent republic. Zālītis wrote of one prominent interwar figure who, "as a talented forest scientist and fanatical forest sector patriot, used his position to introduce and strengthen within forestry that self-sacrificing attitude toward one's responsibilities, one's work in the forest and for the good of the Latvian state, which, having survived several generations, is often evident still today."[63] Zālītis associated Latvia's "centuries-old forestry traditions" not only with scientific advances, but also with "this honorable attitude toward work, toward Latvia's forests as a national treasure, as our popular capital." Aija Zviedre of the Forest Workers' Association concurred:

> In earlier times "ranger" and "state forester" were synonymous with honor. . . . The forest was a matter of honor for every forest ranger. When my father [forest scientist Arvīds Zviedrs] and I visited a ranger, he would show us everything, the little pines and spruces, how beautifully it was all tended. It was a labor of love for him. It was not about fulfilling the plan, it was about responsibility toward the forest.[64]

Indeed, for Zviedre, the forest ranger was the quintessential symbol of a lost golden age, embodying honor, industriousness, knowledge, and sensitivity to nature, as well as a separate and authoritative status vis-à-vis ordinary society. Her childhood reminiscence is deeply nostalgic:

How does the forest ranger appear in my memories? Dressed in a beautiful uniform, his boots shining, with a shotgun over his shoulder, he emerges from the forest thicket and then goes back into it. When he chanced to meet with people, conversation turned to strange happenings—wind-toppled spruces, fawns ravaged by dogs, forest fires—and everyone listened with respect, as if to a visitor from another world, even a "scary" one for women. The forest ranger could impound straying livestock, capture timber thieves, punish tramplers of new plantings, but people did not take offense, for it was all done for the good of the forest. . . . I often encountered ranger Ozols of the Gārša district near Gulbene. Tall and gray-haired, he himself like an oak[65] had become a part of his forest, and it is doubtful whether any natural scientist could exceed the old man's wisdom. He had a story to tell about every stand, every turn in the road, and every phenomenon had its explanation—not from a book, but from his own experience, for this life in the forest had been led with watchful eyes and a sensitive heart. . . . I never saw a ranger loafing about the house during the daytime or doing woman's work. . . . There were among the rangers, of course, some loafers, drunkards or otherwise worthless people, but they did not keep their jobs for long. And their bad reputations preceded them. Public censure punished them more severely than the courts. Honor was the most highly valued trait.[66]

Traditionalists prided themselves in having kept alive, in the face of Soviet oppression, this honorable legacy of vigilantly defending the forests as a national treasure. Veterans of the Soviet period frequently pointed out that forestry was always "the most Latvian" sector of the economy: "The only sector that remained Latvian after the Russian occupation in '40 was forestry. We spoke Latvian, our official documents were written in Latvian." The implicit patriotism of Soviet Latvian foresters was expressed not only through the "perfect, almost military" protection of forests from fire, pests, and other calamities, according to former minister Vītols, but also through an impressive commitment to protecting nature. During the Soviet period, he claimed, "much more was done in the sphere of nature protection than now and in all other years put together." Most of all, though, it was expressed through efforts to maintain and increase Latvia's productive forest reserves. "In the days of the [Soviet] empire," declared Zālītis, "the entire forest leadership self-sacrificingly, fanatically acted to prevent the impoverishment of Latvia's forests." Veterans often recounted with pride that forestry officials had routinely manipulated data submitted to central planners in Moscow in order to reduce the logging quotas imposed on Latvia—not for the sake of reducing their own workloads, as the skeptics would have it, but to prevent overharvesting of Latvian resources.

From this perspective, present-day state foresters and forestry practices were validated by their unbroken links to this patriotic past. The challenge of the post-

Soviet period was to carry on the noble tradition, refurbishing the honor of the state forestry profession where it had been tarnished by Soviet misrule and defending it against the new pressures of post-Soviet capitalism:

> After the [second world] war [and subsequent deportations], only one-tenth of the highest qualified specialists and head rangers, as well as forestry instructors, remained in Latvia. . . . The remaining foresters preserved the enthusiasm and work ethic inculcated by our organizations and forest schools . . . but it cannot be denied that, with generational change, a different, more "contemporary" attitude spread toward work and toward the values entrusted to us. Our task is to help regain the forester's honor and prestige, to become once again true stewards [saimnieki] of our homeland's forests.[67]

The WWF coalition's critique of traditional forestry practices as ecologically and economically irrational was seen as an attack on the "honor and prestige" of Latvian foresters as good stewards of the nation's riches, and quite possibly a deliberate attempt to tarnish this legacy in the eyes of the world. "'Sustainability, biodiversity,' and again, 'biodiversity, sustainability,'" lamented Zālītis; "it sounds modern, but already tiresome. As if the only goal of earlier generations of foresters had been destruction of the forests."[68] Some of the green critics were sincere, claimed Dāboliņš of the Forest Inventory Institute, but others were "terrific demagogues with a great ability to get through even to the world media, in order to tar the accomplishments of Latvia's foresters."[69] For Māris Daugavietis, a forestry professor, recent publications on biodiversity and sustainable forestry had prompted unpleasant questions on the classic Marxist theme of cui bono: "Why are the Latvian forest sector's prestige and competitiveness in international markets being intentionally undermined? Who is financing this campaign? Whom does it benefit?"[70] At its 1997 congress, the Forest Workers' Association formally condemned WWF and Rotbergs for "publicly stating that Latvian forestry is ecologically backward. . . . This claim is in conflict with the fundamental principles, developed through centuries of scholarly and practical work, of Latvian forest cultivation."[71]

Against the WWF critique, traditionalists often mounted the argument that, far from decreasing the forests' ecological value, Soviet Latvian foresters had contributed in important ways to global environmental quality by increasing Latvia's total forest acreage and thereby sequestration of harmful greenhouse gases. They also invoked Latvia's high levels of biodiversity relative to Western Europe as prima facie evidence of good stewardship and suggested that emphasizing the threats to biodiversity was unpatriotic. Thus Jurģis Jansons, a recent forestry graduate and Zālītis disciple, declared ominously: "Latvia's present forest diversity should be advertised as a contrast to Western Europe's truly overurbanized nature. The denial and calling into question of this diversity, and the search for problems where none exist, brings to mind much deeper reasons than . . . personal ambition."[72]

In Praise of the Ulmanis Days

A corollary to the traditionalists' valorization of the state forester was their belief in the state as the only competent and trustworthy long-term steward of public resources and, inversely, their rejection of WWF's faith in markets and private enterprise. As a Forest Service official explained to me, a logger is "like a combine driver hired by a farmer to plow a field": he simply cuts the trees, takes the timber and leaves; by definition, then, he cannot have an economic interest in the long-term well-being of the forest.[73] Traditionalists claimed that, except for one or two subsidiaries of large Nordic concerns, private logging firms in Latvia were mostly "suitcase firms": fly-by-night operations interested only in short-term gain, and therefore not to be trusted with long-term leases or with greater responsibility and freedom in managing the full forestry cycle. The large multinational firms, for their part, were promoting bogus pseudo-ecological claims in order to cut corners at the expense of future forest productivity.

In their reading of interwar Latvia as a golden age, many traditionalists held in particularly high esteem the quasi-authoritarian dictator Kārlis Ulmanis, "highest protector" of Forest Days, with his xenophobic economic nationalism and dirigiste controls over private enterprise. Thus former minister Vītols noted approvingly,

> [Ulmanis] used state power to nationalize private enterprises, even before the Soviet takeover in '40. . . . And Kārlis Ulmanis did this for one reason only: all of these enterprises were controlled by non-Latvians. Non-Latvians, whether they were Russians, or they were Jews, or they were Germans, but in any case they were non-Latvians. For the Latvian had just then let go of his plow and come to Riga, just then traded his farm boots for city shoes.

While applauding Ulmanis's antiforeign expropriations, traditionalists simultaneously celebrated the Ulmanis days as a time when the sanctity of private property was respected by both the state and ordinary citizens. In those days, they maintained, not only was the state strong and state foresters honorable and esteemed, but individuals were worthy and competent property owners. Back then, as Forest Service official Jānis Osis put it, "property was sacred, a family thing: people thought about the benefits to their children and children's children."[74]

In fact, contemporary accounts in the foresters' journal, *Forest Life*, suggest that new landowners were typically as cash-strapped and inexperienced in the 1920s and 1930s as in the 1990s, and that conflicts regarding overharvesting between state foresters and timber-hungry homesteaders were legion. Nonetheless, traditionalists imagined an ideal Ulmanis-era landowner as a foil for today's lapsed landowners. In the post-Soviet era, lamented Osis, "we lack traditions, moral values." Osis's colleague Uldis Grava concurred: "Socialism has damaged the individual. He is not ready to take on the responsibilities that accompany property rights." According to Zālītis, "The last ten years have been the most destructive in the entire 120 years of Latvian forestry," because private forest owners' "only goal is logging and reaping a profit by breakfast time tomorrow."

Traditionalists identified the inexperience and poverty-induced short-termism of new private forest owners as the gravest threat to Latvia's forest resources, and they envisioned a solution not through education and support, as per the WWF agenda, but rather tighter state control. Owners had to be made to understand that the forest was "not only their property, but the state's, the nation's wealth," as chief forester Dišlers put it. "The forest is not private property," agreed Vazdiķis of the Forest Inventory Institute. "The private owner must be to some extent enfettered, freedom must be earned." "We are always saying . . . that private property yields greater productivity, efficiency, that it's always private property *über alles*, but I'm not fully convinced of that," said former minister Vītols. In this view, the state's obligation was not only to prevent overharvesting in private forests, but also to prevent *under*harvesting: that is, to enforce maximum production. "It is in the state's interests for each cultivated hectare to produce," asserted Grava. "Let the landowner produce whatever he wants, but the property must produce." If a landowner failed to maximize his productive potential, the state should fine him or even confiscate the land, Grava maintained, echoing the patriotic homesteader-hero of Jānis Jaunsudrabiņš's interwar agrarian classic *The Homesteader and the Devil*, who militantly declared: "If I were in charge, I would order that poorly run homesteads be transferred to the state land fund, and their owners be declared enemies of the state."[75]

In short, despite assertions about the sanctity of private property, the traditionalists' vision was faithful to the Ulmanis regime's actual legacy of statism. Under Ulmanis's authoritarian rule, the number of state-owned enterprises and monopolies increased constantly, almost entirely through nationalization of private enterprises. Nationalization was justified on the grounds of economic nationalism—of opening up manufacturing and industry, as Ulmanis put it, "to the Latvian plowman and his sons and brothers in the cities."[76] But many of the takeovers were, in fact, of successful Latvian-owned firms. And while the agrarian reform clearly made landed property one of the pillars of the Latvian nation-state, according to Arnolds Aizsilnieks, from the very outset it was not landless individuals but the state itself that profited most from the agrarian reform.[77] Even by the end of the interwar period, the state still owned over half the land in the State Land Fund, including 80 percent of Latvia's total forest lands. In Aizsilnieks's view, increasing state land ownership was a paramount, if unstated, goal of the agrarian reform.

Moreover, the celebration of private landownership by no means implied faith in private enterprise and markets. As Ulmanis himself made clear, the valorization of the family farm as a reservoir of national identity militated against treating it as private property in the capitalist sense: "Our landownership is not and must not be a capitalist enterprise. The farmhouse is the dwelling place of generations of farm families and the eternal source of the nation's living strength. Therefore we must prevent the transformation of farmland and houses into a market commodity and object of speculation."[78] During the era of formal central planning after the 1934 coup, the state assumed full responsibility for determining, as Ulmanis put it, "who will produce what and how."[79] The state directly commanded farmers through the Chamber of Agriculture, and many agricultural support policies were explicitly aimed at circumventing market forces.[80] According to Aizsilnieks, ever-expanding licensing requirements "opened the gates wide for arbitrariness and corruption within the civil service."[81]

It was this legacy of statism, xenophobia, and agrarian productivism that traditionalists embraced as the template for the future development of Latvian forestry. Along with their rejection of the "Swedish methods," they called for increasing the proportion of state-owned forests and for strengthening state control over private enterprise—not only over logging firms and foreign-owned concerns, but also over Latvia's degraded private forest owners, who were seen as having lost the stewardship ethic that prevailed in the interwar era. To this end, they argued for preserving the traditional role and status of state foresters and vehemently opposed the WWF coalition's proposal to reduce the number of rangers in the field. The lynchpin of Latvian forest stewardship was the ranger,

> a person responsible for a specific forest territory, whether it was 400 to 500 hectares as in the thirties, or today's 1,500 to 2,000 hectares, or in the manorial era it was maybe very small—100 to 150 hectares. But always the structure has been created such that at the bottom is a person who knows the area, who has perfectly planted the seedlings and has seen almost in his lifetime how they grow to the first thinning, when for the first time a commercial product can be taken from the forest.[82]

For Aija Zviedre of the Forest Employees' Association, the forest ranger was "the soul of the local community" and the only person who "really knows what's going on" in his ranger district. "Being a forest ranger is not a profession but a way of life," she told me, and the crucial issue was where he lived. "If he lives in the forest, then he is a steward [saimnieks]," but if he lives somewhere in town, then he "becomes like any other paltry civil servant." Likewise, if he takes a job with a private firm, "he is transformed from a saimnieks to a servant." [83] In stark contrast to the WWF coalition's critique of the forest bureaucracy as corrupt and overweening, Zviedre complained that the ranger was not exalted enough, that the state lacked the power to catch and punish timber thieves and keep private forest owners and loggers in line. Forest scientist Miervaldis Bušs concurred: the problem was not that the forest was overrun with too many civil servants, but that "the forest has been emptied of people. Those remaining have moved to the towns. There is no longer anyone who can be called a forest saimnieks."[84]

Globalizing the "Normal Forest"

By the end of the 1990s, WWF had more or less vanquished its traditionalist foes, largely because of the crucial counterhegemonic site it had created at Mežole. Like the "new European wilderness" at Lake Pape, Mežole served a sort of anti-Indrāni landscape, where WWF had inscribed its critique of agrarianism in nature by transforming the parklike normal forest into natural forest. Mežole veterans reported that many of the project's staunchest opponents had become markedly less hostile over the years. In 1999, a Mežole district state forester could confidently report: "No one debates or questions any longer the things we argued about three or four years ago." Notions of biodiversity and natural forests had become unavoidable agenda items in Latvian forest policy debates. In 1998, the traditionalist journal Forest Life launched a regular column on biological diversity. The first installment, by Viesturs Lārmanis, a scientist from the Latvian Fund for Nature, presented the tenets of contemporary forest ecology. Noting that "carefully tended stands are not well suited for the survival of many animal species," Lārmanis declared heretically, "In this century, Latvia's forests were in the worst shape during the first period of independence. This was the era of lovely, well-tended, and often also overlogged forests."[85]

Responding to lobbying by the WWF Forest Club, Latvia's liberal prime minister reorganized the Forest Service in June 1997 and appointed several WWF allies to leadership positions: Arnis Melnis, a Forest Club member and executive at the Latvian-Finnish joint venture Silva, was named director; Mežole district ranger

Aigars Dudelis was made deputy director; and project director Otto Žvagiņš was appointed head of the Division of Development, Support, and Information. The reconstituted Forest Service promptly adopted some biodiversity-friendly norms in logging regulations: requiring dead trees and old trees to be left in cuts, for example, and permitting logging based on natural boundaries, though only within the confines of the standard geometric logging compartment. The principles of sustainable forestry were embraced in a national Forest Policy approved in April 1998, which called for conserving biodiversity through management for natural processes. The Forest Service distributed WWF's forest ecology handbook to all forestry districts, incorporated GIS technology into proposals for reorganizing Latvia's forest information system, and began developing its own GIS pilot project. A Swedish-sponsored project was launched in 1997 to identify key habitats in all of Latvia's forests, using the Mežole ecological inventory methodology. At the end of 1998, the Mežole Project itself was taken over by the Forest Service, which began developing site-adapted harvesting plans for the project territory, thanks to exemptions from classic clear-cut regulations secured by WWF. In 1997, Latvia became the first East European country to begin developing national forest certification standards within the Forest Stewardship Council process.[86] According to the FSC task force coordinator, the role of Mežole as a demonstration site and "a place for all the big shots to go and argue" was critical in moving the drafting process along; without it, he said, "we in the certification group would have spent a lot of time talking about things we have never seen, but only heard about from international practice."[87]

Most important, in 1998 the Forest Service launched a project, with funding and technical assistance from the United Nations Food and Agriculture Organization (FAO), to analyze the forest sector and prepare proposals for a fundamental restructuring. The reforms enacted as a result of the FAO project were largely in line with the recommendations of the WWF Forest Club and the 1995 Swedish-implemented forest sector review. A new state-owned corporation, established on January 1, 2000, assumed ownership and management of state forests. Uģis Rotbergs, only recently denounced as a public enemy by the forestry establishment, was appointed to the board of directors. The Forest Service's mandate was restricted to enforcing forestry regulations and supporting private forest owners. The number of forest districts was reduced and the number of state foresters and rangers nearly halved. Legalizing licensed private providers eliminated the state monopoly in forest management planning. According to the FAO project director, its overriding aim was "finally to end state intervention in the market with all manner of rules and documents and so forth."[88] The FAO reform heralded the beginning of the end of the era of the management directive and of the "civil servant as God."

In short, the WWF campaign succeeded to a very significant degree in reinventing the Latvian forest sector in line with Western scientific narratives and market liberalism. Despite these dramatic ideological and policy shifts, however, many traditionalists remained hostile to the reforms. They denounced the diminution of the Forest Service's powers and warned that transfer of ownership to a corporation might pave the way for eventual privatization of state forests. "I would like everything to stay in the hands of one *saimnieks,*" lamented Aija Zviedre. "The forest in the hands of one *saimnieks.* I seed it, I tend it, I sell it. . . . I sell what I have grown, it is mine. The result of my labor. The result of the labor of my state institution. Why can't it be like that?" Yet Zviedre plaintively acknowledged that hers was "apparently an old-fashioned and unpopular view" and that the FAO reform was "irreversible. It's a different world now and things will never be as they once were, for the rangers no longer live in the forest. . . . A hired hand is a hired hand, and a *saimnieks* is a *saimnieks,* but we will probably never go back." The traditionalist critique was voiced not only by veteran foresters but by outsiders as well. As late as 1999 biologist Jānis Priednieks of the Latvian Fund for Nature expressed "deep skepticism" about long-term logging leases, criticized the Mežole Project as a front for logging firms, denounced on ecological grounds the principle of logging along natural boundaries, and identified private forest owners as the greatest threat to conservation.[89] The president of the Environmental Protection Club, Latvia's leading informal green group, lamented the loss of the "*saimnieks* consciousness" among private forest owners and maintained that "the system of state repression must be used against incorrect forest management."[90]

Contesting Forest Science: "Latvia Is Not the Earth's Navel"

The scientific foundations of forest management remained a critical axis of conflict between reformers and traditionalists. At one level, the battle was pitched between competing interpretations of Latvian forestry traditions. While traditionalists invoked the writings of interwar Latvian foresters to defend current practices such as drainage and monoculture planting, Rotbergs scoured the same sources to find evidence of what he saw as incipient recognition of ecological principles. He kept a file of passages that defined the forest as a "deeply interconnected whole" comprising a multitude of animal and plant species; urged Latvian foresters to "know and understand the character of the virgin forest and derive management principles and practices from it"; argued against removing cutting remains and maintaining a "clean," parklike forest; and advocated mixed-stand regeneration to "protect the forest's balance in future centuries."[91] Rotbergs recalled that as a graduate student at the Silva Forest Science Institute in the early 1980s, "I began to collect all sorts of articles from [interwar issues of] *Forest Life* and so forth, because it started to bother me that people invoked other people's names and said—earlier it was like this. None of them had actually read it . . . and the quotations were condensed and paraphrased in an annoying way."[92] In 1999, WWF published a revisionist project, *Latvian Forest History through 1940*, that explored at length the dynamics of forest cover and composition, animal life, forest policy and legislation, timber trade, and forest exploitation. With this volume, Rotbergs hoped to broaden and deepen historical awareness of Latvian forests and foresters beyond the temporal and conceptual scope of the German normal forest.[93]

At another level, the battle over forest management reflected a deep divide between inward- and outward-looking conceptions of science and the pursuit of knowledge. For WWF and its allies, scientific progress was a product of the international flow of ideas. Rotbergs recalled how as a graduate student he was drawn to explore not only Latvian classics but also the foreign literature available in the institute's library: "They had great Scandinavian, Canadian journals there. . . . I read, say, all the IUFRO [International Union of Forest Research Organizations] congress reports that seemed interesting. . . . And I sat and sat there, until the scientific council expelled me on the grounds that I read too much!" In contrast, Rotbergs saw mainstream Latvian forest science as backward and parochial. Forest ecologist Normunds Priedītis agreed: Latvians were still debating fundamental ecological principles accepted by "the majority of the world's countries," he lamented, in large part because "we have never received in our libraries the leading international journals in ecology and vegetation science. Because in Soviet times the dominant sciences here were biochemistry, microbiology, pharmaceutical manufacturing. . . . And so the only alternative is to get some kind of scholarship or international award and try to break out somewhere to the outside."[94]

Priedītis himself was able to "break out" in the early 1990s when, as a WWF project executant, he was awarded a Prince Bernhard Scholarship to study at Sweden's Uppsala University, where he completed doctoral research on the ecology of wet forests. He continues to keep abreast of current developments through membership in international associations, going to international conferences, and subscribing to as many professional journals as he can afford on his modest income. "At least once a year or so when traveling to symposiums and other events," he told me, "with great difficulty and in great haste I manage to get to the libraries and go through the journals in my field, . . . and by copying the necessary articles or ordering copies from the authors, it is possible fairly normally to follow current global developments in ecology." He was at pains to identify with geographical precision the international sources of validation of his own research:

> I have to say that my only publication in Latvian is the book recently published by WWF. Absolutely all the rest of it is published solely in international journals . . . : the Gustav Fischer Verlag journal, *Flora*, two articles in *Plant Ecology*, by Kluwer Academic, a piece just now accepted in the English journal *Environmental Conservation*, published by Cambridge University Press. Then the journal of the Finnish Academy of sciences, *Annales Botanici Fennici*, and the journal of the Polish [*sic*] Academy of Botany [*Folia Geobotanica & Phytotaxonomica*].

In a sense, Priedītis and Rotbergs were answering the call of the great internationalizer Krišjānis Valdemārs, who asked rhetorically in 1888: "Are natural sciences necessary for the majority of Latvian readers?" Decrying Latvians' widespread ignorance of, even hostility to, foreign scientific ideas and words, Valdemārs raised funds among his fellow Latvian students in Moscow to commission popular articles in Latvian explaining these new advances. "First it must be said that new knowledge cannot be described with old Latvian, German, or Russian words," he declared. "New words must be created for new knowledge, things, professions, and ideas." The effort of struggling with new words and ideas would be well worth the while, he assured his countrymen, for "ultimately the reader will be greatly pleased that now he understands everything and sees nature with her wonders in an entirely new light."[95]

Pēteris Zālītis, on the other hand, continued to define scientific validity by looking inward and backward, rejecting as foreign and unnecessary both the substance and the terminology of contemporary forest ecology and conservation biology. "The Latvian language contains all terms related to forests and forestry," he insisted. "I believe we have no need of the term 'key habitats' that is now invoked so often in our press simply because it is a translation, albeit a poor one, from English."[96] This attitude was shared not only by other Soviet-era eminences grises, but by younger foresters as well. In 1998 Zālītis's student Jurģis Jansons published a polemical attack on the WWF approach in *Forest Life*'s new column on biodiversity. Jansons argued that because of its brief tenure, the Mežole Project had generated no longitudinal data "that would allow for correcting or changing the centuries-old experience of Latvian forestry, upon which today's forest management system is based." He questioned the utility of the woodland key habitats methodology and asserted the ecological adequacy of the traditional Latvian system of forest classification. Even further, Jansons questioned the utility of the very notion of biodiversity itself, arguing, "The scientific evaluation of biological diversity is made difficult or even impossible by the lack of clarity worldwide not only concerning its measurement, but even its very definition."[97]

Jansons's piece elicited an equally vitriolic rebuttal from Priedītis, who expressed dismay at such apparent ignorance of and disregard for internationally accepted scientific findings. Priedītis enjoined Latvians to broaden their understanding of forest ecology by studying world literature:

> The ability to communicate with the entire world and the universal availability of literature are the most fundamental gains in the short history of the second period of [Latvian] independence. After carefully following what others have already achieved in studies of ecology and species and population biology, it is possible that not all of the findings of our local "classics" will turn out to be quite so classic. Latvia is most likely not the earth's navel, where ecological processes, including the disappearance of forest habitats and species, follow some unique laws. . . . Natural sciences are a sphere where humanity's knowledge is growing particularly rapidly. It would be very interesting to learn where in this centuries-old experience is the place for the theories of fragmentation, island biogeography, disturbance dynamics . . . and many other principles identified only in the 1960s–1990s?[98]

The invitation for young people to read books in the English language is undeniably commendable," Jansons remarked acidly in his counter-rebuttal. "Such reading broadens people's horizons and stimulates personal development. But along with book reading, one should also promote spending time in the 'world of nature,' which in Latvia's conditions often does not correspond to descriptions in foreign books." Noting the radical impoverishment of nature in Western Europe, Jansons declared: "Although ecological laws are the same everywhere, theories of natural diversity conservation originating in countries characterized by such 'habitats' *have no place in Latvia.*" Jansons went so far as to compare contemporary conservation biology theories to the Stalin era's Lysenkoism, "when energetic people denied the findings of classical genetics and announced a radical change of direction for the field."[99]

For traditionalists, in short, scientific knowledge was the product of a unique national history and environment, while for internationalists it was meaningful only in a global context. Science was thus another arena of competition between Latvians' dueling developmental visions. For internationalists like Priedītis, scientific knowledge was another asset that Latvians must learn to trade on the global market as they made their way in the world:

You could ask Professor Zālītis in which of the accepted international scientific journals—here I mean those that might be published by Gustav Fischer Verlag, by Oxford University Press, by Cambridge University Press, by Springer Verlag—in which of these publications he has published his results. Because each of us can present ourselves and prove whatever we want, but there is some kind of threshold, some threshold of acceptance from the scientific world, which we must surmount in order to declare—yes, this study I have conducted truly is what I say it is. If a scientist believes something on his own and he publishes it in the proceedings of his scientific institute, in his native language, and after that he seeks to maintain that he has overturned everything that up to now has been known to the science of ecology, then, of course, that is laughable. This is truly a problem here in Latvia. . . . I have always believed that the people who come five, ten years after me—they should definitely be smarter than I am and go faster and farther and so on. But unfortunately even from this very newest generation, who are to a certain degree the students of these orthodox scientists, I have encountered such shocking ideas that I cannot understand what these young people will do five or ten years from now, fifteen years from now, in the global scientific community.

For Priedītis, the traditionalist scientific paradigm, like the agrarian attachment to the farmstead, was a developmental dead end: "Lately we often speak of entering Europe. It seems that science, too, should go along."[100] But for traditionalists, sustainable forestry was another stalking horse for a sinister and threatening Europeanization. "A specter is haunting Latvia—the specter of the European Union," warned Zālītis. "I cannot agree with the formulation of the European Union as an end in itself or a 'kingdom of heaven.'"[101]

The clash between resourcist and ecological understandings of forests is common to many societies, but in Latvia, the particular contours of this clash have been shaped by the internationalist-agrarian divide over what it means for Latvians to be good stewards of their land in the twenty-first century. For agrarian traditionalists, this meant cultivating the forest ethnoscape in keeping with the primordial aesthetic sensibilities of the Latvian nation and with the scientific traditions and patriotic values of German-schooled foresters from the interwar years. For internationalists, the *saimnieks* can responsibly steward the forest's natural and economic riches only by participating in the cross-border flow of markets and ideas. In this counterhegemonic vision, the forest remains one of the nation's "chief treasures" and developmental assets, though no longer as parklike ethnoscape, but rather as a source of biodiversity and internationally certified green timber.

At Lake Pape and at Gauja National Park, agrarian nationalists to some degree succeeded in keeping their ethnoscape while globalizing it too. By participating in various Western-supported projects and embracing the internationalist narratives of biodiversity and sustainability, they strengthened their efforts to protect the human-shaped agrarian landscape. Traditional foresters, in contrast, like the radical productivists of northern Kurzeme, rejected the internationalist agenda outright and remained largely aloof from globalizing sites like the Mežole Project. The potent mythology of the forest sector, with its notion of a special brotherhood united and set apart from civilian society by patriotic honor and commitment to "centuries-old traditions," proved exceptionally resistant to colonization by Western discourses.

Yet unlike what occurred in northern Kurzeme, traditional foresters failed to stall the globalization of their ethnoscape. This was partly owing to WWF-Latvia's lobbying efforts, which were considerably more effective than those of the northern Kurzeme park boosters, but undoubtedly in larger part because of WWF's alliance with domestic and transnational corporate actors and the far greater economic interests involved. Advocates of biodiversity gained greater clout, not surprisingly, when what was at stake was not simply the hazy prospect of future ecotourism, but the very real dynamics of a vital export sector. Yet while the agrarian traditionalists' policy agenda was defeated at the level of institutional reform, their oppositional discourse remained alive and well in science, academe, and the popular press. At the end of the first post-Communist decade, globalization had not yet succeeded in teaching all Latvians to see the forest through Western eyes.

Conclusion

In April 2004, on the eve of Latvia's accession to the EU, WWF-Latvia's director Uģis Rotbergs offered these cautiously hopeful reflections in an interview:

> Latvia brings a wealth of nature, a richness in biodiversity that has disappeared from most parts of Western Europe. But I also think that we bring an opportunity to re-think development, to do things differently, and hopefully better. This of course does not just apply to Latvia, but to all the countries of Central and Eastern Europe, with their different living standards, cultural values, and relatively intact natural environment. . . . The real challenge for Central and Eastern European countries is to create national visions—how we want to live, what we want to do, what we want our environment to be like in future. Our accession to the EU has opened up many opportunities to develop, to move forward. But we have missed many, or not made full use of them. . . . If anything, Latvia's accession to the EU on May 1 does not mark the arrival at a destination, but much rather the start of a longer journey—not only for us Latvians and other Central and Eastern Europeans, but for all European citizens.[1]

The particular paths traveled on this "return to Europe" will ultimately determine the fate of biodiversity on the European continent. EU membership may allow the post-Communist countries to avoid the environmental mistakes of the industrialized West, as Rotbergs hopes, or doom them to repeat those mistakes in the pursuit of Western-style development. Brussels may rise to the challenge of responsibly stewarding its natural windfall, in keeping with the official discourse of sustainability, integrating environmental protection into sectoral policy, and "halting biodiversity loss by 2010." Or familiar trends may be reproduced by privileging perverse subsidies and destructive infrastructure projects.

The journey will also help determine the fate of each post-Communist society as a nation, as a community looking simultaneously backward and forward as it seeks to control its destiny. Having entered the bureaucratic labyrinths of another supranational union, to what extent will new members be able to shape the conditions of their membership? For Latvians and other agrarian-identified East Europeans, will it be possible to negotiate the "return to Europe" while carrying the

weight of the "nation of farmers"? Can agrarian nationalists save their ethnoscape while globalizing it too, or will the reimagining of nature as biodiversity
necessarily erase indigenous landscapes of culture and labor?

In Latvia's rural communities and in professional rural development and conservation circles, agrarian responses prevailed during the 1990s. Western initiatives were largely welcomed when they supported preserving the ethnoscape as a
cultivated landscape of labor and were resisted when they did not. But while internationalists may have been outnumbered by agrarians, they nonetheless had the
resources and influence to mount powerful counterhegemonic efforts. And when
internationalists were backed up by international capital—as in the case of WWF-
Latvia and its Nordic timber allies in the sustainable forestry campaign—these efforts succeeded in overturning deeply rooted agrarian orthodoxies. In the coming
years, as membership in the EU opens the door much wider to international capital in the form of foreign investment, development loans, and EU subsidies, local
actors will likely find it increasingly difficult to resist global norms.

Some readers will doubtless wonder whether material interests play a larger
role in the case studies reported here than the national discourses to which I give
explanatory power. Perhaps so-called agrarians are resisting Western models of
nature management simply because these models threaten their livelihoods or
bases of political power, rather than because of some historically rooted attachment to the nation of farmers and its ethnoscape. Interests are not irrelevant to
this story, but interests alone cannot explain the breadth and depth of the contestation described here. The role of interests looms largest in the forestry case,
where the economic stakes on both sides are both more substantial and more
clearly defined: the profit margins of multinational timber companies on the one
hand, and the livelihoods and modus operandi of thousands of civil servants and
old-school logging enterprises on the other. But even in this case, not all of the
key actors had obvious material stakes in one outcome or the other. As a young
masters student, Jurģis Jansons could reasonably have envisioned a more lucrative and glamorous future for himself as an internationally recognized scientist, publishing in Western journals and winning scholarships to study and travel
abroad, rather than as a doctrinaire defender of his graying mentor, Pēteris
Zālītis.

In the forestry case as in the others, those whom I have labeled agrarians and
internationalists do not divide neatly along occupational or even generational
lines. Some young forest rangers clung to the security of the old paradigm, while
others imagined a better future for themselves in a forest sector transformed by
international norms and foreign capital. WWF's Uģis Rotbergs and Jānis Pried-

nieks of the Latvian Fund for Nature, both directors of Western-funded conservation organizations, with similar educational backgrounds and roughly the same age, passionately defended radically divergent visions of proper nature stewardship and rural development. Interests cannot tell the whole story because they are themselves constituted by ideas and discourses, particularly in periods of great uncertainty about the future. For a young forester, as for a schoolteacher or ex-kolkhoz worker in a remote rural township during the chaotic 1990s, the surest path to a prosperous and dignified future was by no means self-evident. As Latvians negotiated their historically unprecedented integration into the global market economy and into an unpredictably metamorphosing European Union, the calculation of interests was sketchy at best, and visions of possible or desirable futures for themselves and their communities drew heavily upon the templates provided by national discourses.

In the aftermath of Latvia's accession to the EU, as the quest for Western levels of prosperity kicks into a higher gear, agrarian nationalists will have to ponder more closely what kind of ethnoscape it is possible to preserve within the union. Skeptics are justly concerned about West Europeans seeking to preserve the Eastern countryside as a sort of theme park of quaint villages, horse-drawn wagons, and simple farm folk (with the odd Soviet missile silo adding a bit of cold war flavor). This is precisely what has already happened to many rural places and their inhabitants in Western Europe itself, as well as in more exotic locales. The very attempt to preserve heritage and lifeways, notes Yi-Fu Tuan, often turns its subjects "into figures in glass cases, labeled and categorized as in a museum, for the delectation of world-hopping tourists and scholars."[2] Is an ethnoscape preserved as a tourist attraction still an ethnoscape? In the English case, David Lowenthal suspects it is not: "The heritage landscape is less and less England, more and more 'England-land,' Europe's offshore theme park."[3] Today, the inhabitants of Europe's new "post-Communism-land" confront Tuan's question: "Can a people really be proud of being a museum piece, an endangered cultural type, a song-and-dance attraction for tourists? For lack of a promising future, the answer may well be yes—a 'yes' reinforced by powerful outsiders who tell the locals and natives that they, too, serve science and humankind by standing still, living as they have always done."[4]

Underlying these questions is the more fundamental one posed by Rotbergs: will the citizens of the new member states articulate coherent "national visions" for their future within Europe, or will they be passive objects of integration processes imposed from the outside? Moreover, to what extent are nationally defined developmental visions still relevant in a supranational Europe and a globalized economy? Theorists in the instrumentalist-modernization mold have long argued that national identity is a waning force in the new world order. As Eric Hobsbawm puts it, "Nationalism, however inescapable, is simply no longer the historical force it was."[5] Present-day nationalism offers no unifying "positive program," he argues, but only xenophobia and unrealistic Mazzinianism, or the bankrupt belief that the borders of nation and state should be congruent. Especially for "small linguistic communities vulnerable to quite modest demographic changes," such as the Welsh, Estonians, or Latvians, national identity is only "a cry of anguish or fury. . . . Time and again such movements of ethnic identity seem to be reactions of weakness and fear, attempts to erect barricades to keep at bay the forces of the modern world." Nationalism represents an unfortunate substitute for a social fabric torn by the dislocations of globalization and social atomization, Hobsbawm maintains.[6] In contrast to fundamentalism, for example, the "call of ethnicity or language provides no guidance to the future at all. . . . It is merely a protest against the status quo" or "the others." For Hobsbawm, in short, the future is largely supranational and infranational. While national history and culture will continue to "bulk large—perhaps larger than before—in the educational systems and cultural life" especially of smaller countries, nonetheless "the slogan of self-determination . . . can offer no solution for the twenty-first century."[7]

Against such predictions, I have argued that while national discourse often does express an anguished cry of protest against modernity, at the same time it offers a "positive program" for the future. As Roman Szporluk observes in his comparative analysis of Friedrich List and Karl Marx, in the nineteenth century nationalism and Marxism "were competing theories of industrialization and indeed were rival programs for a modern society."[8] In post-Soviet Latvia, too, nationalism—or more specifically, competing notions of nation and homeland based upon rival readings of national history—structure particular developmental programs. The content of particular national discourses continues to be relevant in explaining how societies negotiate modernity. "It is never sufficient to see in such myths and symbols obsolete compensations for much-feared social change," contends Anthony Smith. "Nostalgia is so often linked with utopia; our blueprints for the future are invariably derived from our experiences of our pasts."[9] In navigating their return to Europe, Latvians' visions of rural development and nature management have been and will continue to be shaped by discourses of nation and homeland.

Many observers have theorized that globalization is not only eroding national identity, but also diminishing the significance of territory itself. In the near future, in this view, "the fundamental ways in which we classify human beings will no longer be identified with territory."[10] Thanks to the growth of cross-border movements, even territorial communalisms are increasingly transnational. In this post-territorial world, David Morley and Kevin Robins argue,

> There can be no recovery of an authentic cultural homeland. In a world that is increasingly characterized by exile, migration and diaspora, with all the consequences of unsettling and hybridization, there can be no place for such absolutism of the pure and authentic. In this world, there is no longer any place like Heimat. More significant, for European cultures and identities now, is the experience of displacement and transition.[11]

The case studies presented in this book support the contrary argument that even as territoriality is being reshaped by the accelerating flows of people and ideas across increasingly porous boundaries, it is not losing its significance.[12] The relationship between internationalism and territoriality is not necessarily dichotomous. Internationalism can be wielded not only to transcend territoriality, but also to reimagine territoriality. From the seafaring visions of Krišjānis Valdemārs through the wilderness restorations of WWF, Latvians have embraced internationalism as an expression of territoriality, not a rejection of it or an alternative. They have embraced openness, transit, and cultural hybridity not to dissolve Latvians into some larger nonterritorial culture, but to make Latvians better stewards of their own land, and thereby more prosperous and respected citizens of the world. An internationalist like WWF's Rotbergs may be a full-fledged member of the e-mailing, jet-setting, global intelligentsia, but he remains a Latvian nationalist, and by no means a "laissez-faire cosmopolitan, who has 'seceded' from the nation-state."[13]

In short, internationalism is not always antiterritorial, and territoriality is not always anticosmopolitan, exclusionary, and othering. Many Latvian proponents of an inward-looking territoriality—defenders of the agrarian ethnoscape—simultaneously look outward by participating in globalizing projects and appropriating Western notions like ecotourism and biodiversity. To borrow Arjun Appadurai's formulation, rural development and nature management policymaking in post-Soviet Latvia have constituted "a space of contestation in which individuals and groups seek to annex the global into their practices of the modern."[14] Not only internationalists like Rotbergs, but also agrarian nationalists like the geographer Aija Melluma or Jānis Priednieks of the Latvian Fund for Nature, have sought to incorporate global notions of biodiversity into their strategies for national development.

While this phenomenon demonstrates the growing scope and penetration of transnational norms and practices, it also shows that national discourses can survive within and adapt to this penetration. It can help to explain the surprising persistence of the "potato principle" in a largely postagrarian and internationalizing Eastern Europe. "In the age of 'transnationalism,'" observes Katherine Verdery, "a concern with nations rooted in the specific soil of their particular states seems oddly anachronistic."[15] But if internationalism can be not the antithesis of, but rather a strategy for national and territorial development visions, then the coexistence of inward-looking and outward-looking identity discourses appears less surprising. In Latvia in the 1990s, the same globalizing forces that raised the specter of biodiversity-oriented neocolonialism also empowered the agrarian resistance. Western environmental aid projects imported the notion of nature as biodiversity, but they also imported the participatory planning mechanisms that gave defenders of the ethnoscape at Gauja National Park an influential role, and the discourse of "stakeholder democracy" made it impossible to establish a Northern Kurzeme National Park against the wishes of the local population. Citizen participation produced ardent advocates of Western-funded rural economic diversification, such as Rucava's Mayor Jānis Veits, but also bolstered hard-line agrarian productivists like Maija Rēriha, mayor of Kolka. Globalization may have framed the agenda, but it did not determine the outcome of the debate.

The Latvian case thus suggests that transnationalism may not entirely erode the meaning of Heimat, although it will certainly leaven the "absolutism of the pure and authentic" with a more outward-looking sense of place. To save the ethnoscape, in other words, it may well be necessary to globalize it too: to keep his homestead, the traditional good farmer may have to become an international ecotourism provider. It remains to be seen, of course, whether the nationally distinctive agrarian landscape can survive this globalization. Ultimately, "restoring a European wilderness" through conservation and ecotourism may not keep more than a handful of Latvians living and working in the countryside. What is certain, though, is that many Latvians today believe that globalizing nature is the only viable mechanism for preventing the depopulation of the countryside—for keeping Latvians firmly rooted, that is, in their national territory.

What will visitors to the Latvian countryside find in ten or twenty years? Will ecotourists wander through the untamed jungle of a borderless "European wilderness," or take in the manicured fields and authentic farmsteads of "Latvialand"? Will rural Latvians live and work on family farms, or clean up after busloads of German tourists at high-rise seaside hotels, or drive tractors on industrial farms owned by a few efficient Danes? As natural and cultural diversity continue to retreat throughout the industrialized world in the face of intensive agriculture, suburban sprawl, and the relentlessly homogenizing consumerism propagated by transnational megacorporations, the answers to these questions should concern not only Latvians and their fellow East Europeans, but indeed all of us.

NOTES

Introduction

1. The new members of the EU are the Czech Republic, Estonia, Hungary, Latvia, Lithuania, Poland, Slovakia, and Slovenia, as well as Cyprus and Malta. Negotiations were initiated in 1999 with Bulgaria and Romania, but these countries have not yet satisfied the EU's conditions for membership.

2. "The vast tracts of relatively unpolluted lands [remain] an invisible backdrop to these striking images" of environmental disaster areas. Andrew Tickle and Ian Welsh, "The 1989 Revolutions and Environmental Politics in Central and Eastern Europe," in *Environment and Society in Eastern Europe*, ed. Tickle and Welsh (Harlow, UK: Longman, 1998), 17.

3. Petr Pavlinek and John Pickles, *Environmental Transitions: Transformation and Ecological Defence in Central and Eastern Europe* (London: Routledge, 2000), 41.

4. See Tim Unwin, Judith Pallot, and Stuart Johnson, "Rural Change and Agriculture," in *East Central Europe and the Former Soviet Union: The Post-Socialist States*, ed. Michael Bradshaw and Alison Stenning (Harlow, UK: Pearson Education, 2004).

5. F. E. Ian Hamilton, "Transformation and Space in Central and Eastern Europe," *Geographical Journal* 165, no. 2 (1999): 135–44.

6. William Cronon, "Introduction: In Search of Nature," in *Uncommon Ground: Rethinking the Human Place in Nature* (New York: W. W. Norton, 1996), 25–26.

7. Anthony D. Smith, *The Ethnic Origins of Nations* (Oxford and New York: Basil Blackwell, 1986), 23.

8. I use this term in Anthony Smith's sense of a physical landscape invested with ethnic meanings, and not in Arjun Appadurai's more metaphorical sense of "global ethnoscapes" as the shifting and deterritorialized "landscapes" created by the movements of "tourists, immigrants, refugees, exiles, guest workers, and other moving groups and individuals." Anthony D, Smith, "Nation and Ethnoscape," in *Myths and Memories of the Nation* (Oxford: Oxford University Press, 1999), 149–59; Arjun Appadurai, *Modernity at Large: Cultural Dimensions of Globalization* (Minneapolis: University of Minnessota Press, 1996), 33.

9. A. Smith, *The Ethnic Origins of Nations*, 186.

10. Simon Schama, *Landscape and Memory* (London: HarperCollins, 1995), 15.

11. David Lowenthal, "European and English Landscapes as National Symbols," in *Geography and National Identity*, ed. David Hooson (Oxford: Blackwell, 1994), 21.

12. Piers Gruffud, "Remaking Wales: Nation-Building and the Geographical Imagination, 1925–50," *Political Geography* 14, no. 3 (1995): 219–39.

13. Carol B. Bardenstein, "Trees, Forests, and the Shaping of Palestinian and Israeli Collective Memory," in *Acts of Memory: Cultural Recall in the Present*, ed. Mieke Bal, Jonathan Crewe, and Leo Spitzer (Hanover, NH: University Press of New England, 1999), 159; Tom Selwyn, "Landscapes of Liberation and Imprisonment: Towards an Anthropology of the Israeli Landscape," in *Anthropology of Landscape: Perspectives on Place and Space*, ed. Eric Hirsch and Michael O'Hanlon (Oxford: Clarendon Press, 1995).

14. See Roderick Nash, *Wilderness and the American Mind*, rev. ed. (New Haven: Yale University Press, 1973); and Alfred Runte, *National Parks: The American Experience* (Lincoln: University of Nebraska Press, 1979).

15. William Cronon, "The Trouble with Wilderness; or, Getting Back to the Wrong Nature," in *Uncommon Ground: Rethinking the Human Place in Nature*, ed. William Cronon (New York: W. W. Norton, 1996); Max Oelschlaeger, *The Idea of Wilderness: From Prehistory to the Age of Ecology* (New Haven: Yale University Press, 1991).

16. Tim Forsyth, *Critical Political Ecology: The Politics of Environmental Science* (London: Routledge, 2003), chap. 2.

17. Raymond L. Bryant, "Power, Knowledge and Political Ecology in the Third World: A Review," *Progress in Physical Geography* 22, no. 1 (1998): 88.

18. Nancy Lee Peluso, *Rich Forests, Poor People: Resource Control and Resistance in Java*. (Berkeley and Los Angeles: University of California Press, 1992).

19. Philip Stott and Sian Sullivan, "Introduction," in *Political Ecology: Science, Myth and Power*, ed. Stott and Sullivan (London, Arnold, 2000), 1.

20. David Takacs, *The Idea of Biodiversity: Philosophies of Paradise* (Baltimore: Johns Hopkins University Press, 1996), 99.

21. Jonathan S. Adams and Thomas O. McShane, *The Myth of Wild Africa: Conservation Without Illusion* (Berkeley and Los Angeles: University of California Press, 1992); Peter R. Wilshusen, Steven R. Brechin, Crystal L. Fortwangler, and Patrick C. West, "Reinventing a Square Wheel: Critique of a Resurgent 'Protection Paradigm' in International Biodiversity Conservation," *Society and Natural Resources* 15, no. 1 (2002): 17–40.

22. James McCarthy, "First World Political Ecology: Lessons from the Wise Use Movement," *Environment and Planning A* 24 (2002): 1286.

23. Eric Darier, "Foucault and the Environment: An Introduction," in *Discourses of the Environment*, ed. Eric Darier (Oxford: Blackwell 1999). See also Arturo Escobar, "Whose

Knowledge, Whose Nature? Biodiversity, Conservation and the Political Ecology of So-cial Movements," *Journal of Political Ecology* 5 (1998): 53–82.

24. Sustainability first emerged as a rethinking of the radically antigrowth position artic-ulated by environmentalists in the 1970s, most notoriously in the Club of Rome's 1972 report: Donella H. Meadows et al., *The Limits to Growth* (London: Earth Island, 1972). The first important articulation of the concept was produced jointly by the quasi-governmen-tal International Union for the Conservation of Nature and Natural Resources (IUCN) and its nongovernmental partner, the World Wide Fund for Nature (WWF), in their *World Conservation Strategy: Living Resource Conservation for Sustainable Development* (Gland, Switzer-land: IUCN, UNEP, WWF, 1980). The UN Environment Program (UNEP) embraced the principle in its influential 1987 Brundtland report: World Commission on Environment and Development, *Our Common Future: Report of the World Commission on Environment and De-velopment* (Oxford: Oxford University Press, 1987).

25. See, for example, World Resources Institute, the World Conservation Union, and the United Nations Environment Programme, *Global Biodiversity Strategy: Guidelines for Action to Save, Study, and Use Earth's Biotic Wealth Sustainably and Equitably* (Washington, DC: WRI, 1992).

26. Dick Richardson, "The Politics of Sustainable Development," in *The Politics of Sustain-able Development: Theory, Policy and Practice Within the European Union*, ed. Susan Baker, Maria Kousis, Dick Richardson, and Stephen Young (London: Routledge, 1997), 43.

27. Darier, *Foucault and the Environment*; Arturo Escobar, "Constructing Nature: Elements for a Poststructural Political Ecology," in *Liberation Ecologies: Environment, Development, So-cial Movements*, ed. Richard Peet and Michael Watts (London and New York: Routledge 1996); Michael Redclift, *Sustainable Development: Exploring the Contradictions* (London and New York: Methuen 1987); Wolfgang Sachs, ed., *Global Ecology: A New Arena of Political Conflict* (London: Zed Books 1995).

28. See, for example, Barbara A. Cellarius and Caedmon Staddon, "Environmental Non-governmental Organizations, Civil Society, and Democratization in Bulgaria," *East Euro-pean Politics and Societies* 16, no. 1 (2002): 182–222; Stuart Franklin, "Bialowieza Forest, Poland: Representation, Myth, and the Politics of Dispossession," *Environment and Plan-ning A* 34 (2002): 1459–85; Pavlinek and Pickles, *Environmental Transitions*; Staddon, "Lo-calities, Natural Resources and Transition in Eastern Europe," *Geographical Journal* 165, no. 2 (1999): 200–08; Tickle and Welsh, *The 1989 Revolutions*, and David Turnock, "Sus-tainable Rural Tourism in the Romanian Carpathians," *Geographical Journal* 165, no. 2 (1999): 192–99.

29. See, for example, Baker et al., *The Politics of Sustainable Development*; Guy Beaufoy, David Baldock, and Julian Clark, *The Nature of Farming: Low Intensity Farming Systems in Nine European Countries* (London, Institute for European Environmental Policy, 1994); Henry Buller, Geoff A. Wilson, and Andreas Holl, *Agri-Environmental Policy in the European Union* (Aldershot, U.K.: Ashgate Publishing, 2000); Hugh Clout, "The European Countryside: Contested Space," in *Modern Europe: Place, Culture and Identity*, ed. Brian J. Graham (Lon-

don: Arnold 1998); Commission of European Communities Directorate General VI/A1, *Towards a Common Agricultural and Rural Policy for Europe: Report of an Expert Group* (Brussels: European Commission, 1997); and David I. McCracken, Eric M. Bignal, and Susan E. Wenlock, eds., *Farming on the Edge: The Nature of Traditional Farmland in Europe* (Peterborough, UK: Joint Nature Conservation Committee, 1995).

30. Stott and Sullivan, *Political Ecology*, 5. See also Raymond L. Bryant and Sinead Bailey, *Third World Political Ecology* (London, Routledge, 1997); and Peet and Watts, *Liberation Ecologies*.

31. Rogers Brubaker, *Nationalism Reframed: Nationhood and the National Question in the New Europe* (Cambridge: Cambridge University Press, 1996); Terry Martin, *The Affirmative Action Empire: Nations and Nationalism in the Soviet Union, 1923–1939* (Ithaca: Cornell University Press, 2001); Yuri Slezkine, "The USSR as a Communal Apartment, or How a Socialist State Promoted Ethnic Particularism," *Slavic Review* 53, no. 2 (1994): 414–52; Katherine Verdery, "Nationalism and National Sentiment in Post-Socialist Romania," *Slavic Review* 52, no. 2 (Summer 1993): 179–203; and Victor Zaslavsky, "Nationalism and Democratic Transition in Post-Communist Societies," *Daedalus* 121, no. 2 (1992): 97–121.

32. Dagmāra Beitnere, "Vai latvieši ir tikai zemnieki?" *Karogs* 11 (2002): 153.

33. Ernest Gellner, "Nationalism in the Vacuum," in *Thinking Theoretically about Soviet Nationalities: History and Comparison in the Study of the USSR*, ed. Alexander J. Motyl (New York: Columbia University Press, 1992), 251.

34. "Europe is set off by quotation marks because it is not, at least in most European parts of the former Eastern bloc and Soviet Union, just a continent. It is also a utopia of sorts in the regional discourse and is spoken of with reverence, suggesting that it has a powerful conceptual as well as geographic significance." Stukuls, *Imagining the Nation*, 131, 134.

35. Vaclav Havel, address to the European Parliament, Strasbourg, February 16, 2000.

36. Nick Coleman, "Exclave's Future Hangs in Air," *Baltic Times*, January 23–29, 2003.

37. Latvian and Estonian identity discourses share much more with each other than with those in Lithuania, which differs from its northern neighbors in its religious identity (strongly Catholic rather than weakly Lutheran) and its history as a former imperial power, rather than a perpetually subject people.

38. Lennart Meri, president of Estonia, speech to the Royal United Services Institute, London, March 3, 1998.

39. Eiki Berg, "Local Resistance, National Identity and Global Swings in Post-Soviet Estonia," *Europe-Asia Studies* 54, no. 1 (2002): 115–16.

40. Peeter Vihalemm, "Changing National Spaces in the Baltic Area," in *Return to the Western World: Cultural and Political Perspectives on the Estonian Post-Communist Transition*, ed. Marju Lauristin and Peeter Vihalemm (Tartu: Tartu University Press, 1997), 132.

41. Merje Feldman, "European Integration and the Discourse of National Identity in Estonia," *National Identities* 3, no. 1 (2001): 5–21. Ultimately, warns Mikko Lagerspetz, "'Europeanness' may turn out to be a politically correct, fashionable version of xenophobia." Lagerspetz, "The Cross of Virgin Mary's Land: A Study in the Construction of Estonia's 'Return to Europe,'" *Finnish Review of East European Studies* 3–4 (1999): 25. See also Eiki Berg and Saima Oras, "Writing Post-Soviet Estonia on to the World Map," *Political Geography* 19 (2000): 601–25; and Martha Merritt, "A Geopolitics of Identity: Drawing the Line between Russia and Estonia," *Nationalities Papers* 28, no. 2 (2000): 243–62.

42. Samuel Huntington, *The Clash of Civilizations and the Remaking of World Order* (New York: Simon & Schuster 1996), 158, quoted in Vihalemm, "Changing National Spaces," 131. Merje Feldman reports that Huntington's controversial book is "among the most cited and revered scholarly works in Estonia. In 1999, the book was translated into Estonian, complete with a foreword by Toomas Hendrik Ilves, Estonia's Minister of Foreign Affairs. On this occasion, Huntington came to Estonia and spoke at a conference together with Ilves and Prime Minister Mart Laar. . . . Huntington is referred to as a 'western expert' and his concept of civilizational conflict is cited not as an idea but as a fact." Feldman, "European Integration," 10. My experience in Latvia suggests that many Latvians share this view.

43. Lauristin, "Contexts of Transition," in Lauristin and Vihalemm, *Return to the Western World*, 37.

44. Feldman, "European Integration," 13.

45. In 1998 only 39% of residents of Riga were ethnic Latvians and in Daugavpils in the southeastern district of Latgale the figure was 14%. Central Statistical Bureau, *Statistical Yearbook of Latvia 1998*, 63, 65.

46. Miervaldis Birze, "Vai Latvijai paredzēts Īrijas variants?" *Diena*, February 17, 1998.

47. Stott and Sullivan, *Political Ecology*, 1.

48. Jeff Chinn and Robert J. Kaiser, *Russians as the New Minority: Ethnicity and Nationalism in the Soviet Successor States* (Boulder: Westview Press, 1996), 18.

49. Jonathan Franzen, *The Corrections* (New York: Farrar, Straus, and Giroux, 2001), 444.

50. Catherine Wanner, *Burden of Dreams: History and Identity in Post-Soviet Ukraine* (University Park: Pennsylvania State University Press, 1998), 203.

51. John A. Armstrong, *Nations before Nationalism* (Chapel Hill: University of North Carolina Press, 1982), 5.

52. See, for example, Aadne Aasland and Tone Flotten, "Ethnicity and Social Exclusion in Estonia and Latvia," *Europe-Asia Studies* 53, no. 7 (2001): 1023–49; Jeff Chinn and Lise A. Truex, "The Question of Citizenship in the Baltics," *Journal of Democracy* 7, no. 1 (1996): 132–47; Geoffrey Evans and Christine Lipsmeyer, "The Democratic Experience in Divided Societies: The Baltic States in Comparative Perspective," *Journal of Baltic Studies* 32,

no. 4 (2001): 379–401; Victor Gray, "Identity and Democracy in the Baltics," *Democratization* 3, no. 2 (1996): 69–91; Mark Jubulis, *Nationalism and Democratic Transition: The Politics of Citizenship and Language in Post-Soviet Latvia* (Lanham, MD: University Press of America, 2001); Priit Järve, "Two Waves of Language Laws in the Baltic States: Changes of Rationale?" *Journal of Baltic Studies* 33, no. 1 (2002): 78–110; Arunas Juska, "Ethno-Political Transformation in the States of the Former USSR," *Ethnic and Racial Studies* 22, no. 3 (1999): 524–53; Robert J. Kaiser, "Ethnoterritorial Conflict in the Former Soviet Union," in *The Challenge of Ethnic Conflict to National and International Order in the 1990s: Geographic Perspectives* (Washington, DC: United States Central Intelligence Agency Conference Report 1995); David D. Laitin, *Identity in Formation: the Russian-Speaking Populations in the Near Abroad* (Ithaca: Cornell University Press, 1998); Laitin, "Three Models of Integration and the Estonian/Russian Reality," *Journal of Baltic Studies* 34, no. 2 (2003): 197–222; Marju Lauristin and Mati Heidmets, eds., *The Challenge of the Russian Minority: Emerging Multicultural Democracy in Estonia* (Tartu: Tartu University Press, 2002); Vello Pettai, "The Games of Ethnopolitics in Latvia," *Post-Soviet Affairs* 12, no. 1 (996): 40–50; Pettai, "Emerging Ethnic Democracy in Estonia and Latvia," presented at the annual meeting of the Association for the Advancement of Baltic Studies, June 1994; Algimantas Prazauskas, "The Influence of Ethnicity on the Foreign Policies of the Western Littoral States," in *National Identity and Ethnicity in Russia and the New States of Eurasia* , ed. Roman Szporluk (Armonk, NY: M. E. Sharpe, 1994); Graham Smith, "The Ethnic Democracy Thesis and the Citizenship Question in Estonia and Latvia," *Nationalities Papers* 24, no. 2 (1996): 199–216; and G. Smith, "When Nations Challenge and Nations Rule: Estonia and Latvia as Ethnic Democracies," *International Politics* 33 (1996): 27–43.

53. Michael Billig, *Banal Nationalism* (London: Sage, 1995), 61.

54. Quoted in Daina Stukuls Eglitis, *Imagining the Nation: History, Modernity, and Revolution in Latvia* (University Park: Pennsylvania State University Press, 2002), 124.

55. Ibid., 124–25.

56. Sidney Plotkin, *Keep Out: The Struggle for Land Use Control* (Berkeley and Los Angeles: University of California Press, 1987), 2.

57. Virtually everyone I interviewed for this project identified as an ethnic Latvian and spoke Latvian as a first language. Russophone political actors did not publicly weigh in on the issues considered here, at least not in the Latvian-language media, and I did not analyze the Russian-language media. Discourses and attitudes about homeland, nature, and rural development among Russian speakers in Latvia would be a fascinating area to study, but it is beyond the scope of this book.

58. A. Smith, *Ethnic Origins of Nations*, 23.

59. Algimantas Prazauskas observes: "Among the western littoral nations [of the former Soviet region], the Latvians and Estonians are least burdened with memories, images, and symbols of the distant past. Like some other small nations of Eastern Europe, they draw inspiration from the fact of their own survival after seven centuries of foreign dom-

ination." Prazauskas, "The Influence of Ethnicity," 161. The notion of "700 years of slavery" is "mythic" in its conflation of foreign domination with "slavery": historians note that while manorialism appeared in the Baltic territories in the thirteenth century, it co-existed with extensive free peasant holdings, and serfdom per se began to spread only in the sixteenth century.

60. Edmunds V. Bunkše, "God, Thine Earth Is Burning: Nature Attitudes and the Latvian Drive for Independence," *GeoJournal* 26, no. 2 (1992): 203. See also Bunkše, "Reality of Rural Landscape Symbolism in the Formation of a Post-Soviet, Postmodern Latvian Identity," *Norsk geografisk Tidsskrift* 53 (1999): 121–38.

61. "With a few notable exceptions, political and sociological studies of nations and nationalism . . . have tended to treat *place* as an insignificant explanatory variable." Robert J. Kaiser, *The Geography of Nationalism in Russia and the USSR* (Princeton: Princeton University Press, 1994), 4.

62. Jan Penrose, "Nations, States and Homelands: Territory and Territoriality in Nationalist Thought," *Nations and Nationalism* 8, no. 3 (2002): 290.

63. David Morley and Kevin Robins, "No Place Like *Heimat*: Images of Home(land) in European Culture," in *Space and Place: Theories of Identity and Location*, ed. Erica Carter, James Donald, and Judith Squires (London: Lawrence & Wishart 1993), 8. Kaiser's notion of national territoriality (in *The Geography of Nationalism*) also identifies a strictly exclusionary role for homeland. See also Graham Smith, Vivien Law, Andrew Wilson, Annette Bohr, and Edward Allworth, *Nation-Building in the Post-Soviet Borderlands: The Politics of National Identities* (Cambridge: Cambridge University Press, 1998), 49.

64. A. Smith, *National Identity* (Reno: University of Nevada Press, 1991), 64.

65. Gellner, *Nations and Nationalism*, 56. "Too much belief in what is patently not so," concurs Eric Hobsbawm in *Nations and Nationalism since 1780*, 2nd ed. (Cambridge: Cambridge University Press, 1990), 12.

66. Donald Horowitz, *Ethnic Groups in Conflict* (Berkeley and Los Angeles: University of California Press, 1985), 56.

67. Schama, *Landscape and Memory*, 15.

68. Katherine Verdery, "Nationalism, Postsocialism, and Space in Eastern Europe," *Social Research* 63, no. 1 (1996): 86. See also Verdery, "The Elasticity of Land: Problems of Property Restitution in Transylvania," *Slavic Review* 53, no. 4 (1994): 1071–1108.

69. Rogers Brubaker, *Citizenship and Nationhood in France and Germany* (Cambridge: Harvard University Press, 1992), 17. For a seminal exposition within the political science literature of the role of ideas in constituting interests, see Alexander Wendt, "Anarchy is what states make of it: the social construction of power politics," *International Organization* 46, no. 2 (1992): 391–441.

70. On how discourse constitutes meanings and "produces interpretive possibilities by making it virtually impossible to think outside of it," see Roxanne Lynn Doty, "Foreign Policy as Social Construction: A Post-Positivist Analysis of U.S. Counterinsurgency Policy in the Philippines," *International Studies Quarterly* 37 (1993): 302.

71. Ibid.

72. See, for example, Eric Hobsbawm and Terence Ranger, eds., *The Invention of Tradition* (Cambridge: Cambridge University Press, 1983).

73. Crawford Young, "The Dialectics of Cultural Pluralism: Concept and Reality," in *The Rising Tide of Cultural Pluralism: The Nation-State at Bay*, ed. Young (Madison: University of Wisconsin Press, 1993), 24.

74. On the loaded and often misleading categories of constructivism and primordialism, see Alexander J. Motyl, *Revolutions, Nations, Empires: Conceptual Limits and Theoretical Possibilities* (New York: Columbia University Press, 1999).

75. A. Smith, *The Nation in History: Historiographical Debates about Ethnicity and Nationalism* (Hanover, NH: University Press of New England, 2000), 55.

76. A. Smith, *Ethnic Origin of Nations*, 207.

77. I largely agree with Mark A. Jubulis, "Nationalism and the Breakup of the Soviet Union: An Ethno-symbolist Critique of Modernist, Constructionist, and Institutionalist Perspectives," presented at the annual meeting of the Association for the Study of Ethnicity and Nationalism, London School of Economics, April 2004.

78. Mark Beissinger, *Nationalist Mobilization and the Collapse of the Soviet State: A Tidal Approach to the Study of Nationalism* (New York: Cambridge University Press, 2001), 18.

79. On the need for discourse analysis and "thick description" in political ecology, see Peet and Watts, *Liberation Ecologies*, 38.

Chapter 1 | By Land or by Sea

1. Anne Applebaum, *Between East and West: Across the Borderlands of Europe* (New York: Random House, 1994), ix.

2. Andrejs Plakans, *The Latvians: A Short History* (Stanford: Hoover Institution Press, 1995), 1–2.

3. Livland, comprising the territories of present-day central-northern Latvia and Estonia, became a Swedish colony. The Polish-Lithuanian Commonwealth ruled Lettgalia in the southeast and indirectly controlled western Latvia through its vassals in the Duchy of Courland and Semigallia.

4. Anders Henriksson, "Riga: Growth, Conflict, and the Limitations of Good Government, 1850–1914," in *The City in Late Imperial Russia*, ed. Michael F. Hamm (Bloomington: Indiana University Press, 1986), 177. See also Henriksson, *The Tsar's Loyal Germans: The*

Riga German Community, Social Change and the Nationality Question, 1855–1905 (Boulder: East European Monographs, 1983).

5. This was not the case in Lettgalia, the region now comprising the Latgale district in eastern Latvia. After centuries of Polish-Lithuanian rule, Lettgalia was incorporated into the Vitebsk province of Russia. Latgale is thus the great exception in Latvian history; its dominant cultural influences are Russian rather than German, Catholic, and Orthodox rather than Protestant.

6. Hobsbawm, *Nations and Nationalism since 1790*, 116.

7. Miroslav Hroch, *Social Preconditions of National Revival in Europe: A Comparative Analysis of the Social Composition of Patriotic Groups among the Smaller European Nations*, trans. Ben Fowkes (Cambridge: Cambridge University Press, 1985), 8.

8. Herder was a teacher at the Cathedral School and a popular preacher at two principal churches in Riga from 1764 to 1769. According to Herder scholar A. Gillies, "The Riga years were perhaps the happiest of Herder's unhappy life. . . . It was in Riga that he first became alive to the meaning of patriotism, for the public spirit of the self-governing Hanseatic city was a revelation to one who had shrunk from the state-system of Frederick the Great." Gillies, *Herder* (Oxford: Basil Blackwell, 1945), 14.

9. Language, according to Herder, is a people's "most distinctive and sacred possession," as it creates a sense of community and a consciousness of difference from others. Folklore is "the invisible, hidden medium" through which "we actively establish a continuum between ourselves and those that follow upon us." Herder's theory of language as the basic condition of nationhood had a profound impact not only in Latvia, but throughout Eastern Europe, where his "ideas caused the greatest political stir," leading to "the prodigious philological research which accompanied nationalist agitation." F. M. Barnard, *Herder's Social and Political Thought: From Enlightenment to Nationalism* (Oxford: Clarendon Press, 1965), 58, 62, 171–72.

10. Benedict Anderson, *Imagined Communities: Reflections on the Origin and Spread of Nationalism* (London: Verso, 1983), 71. According to Anthony Smith, cultivating identification with the homeland is particularly important for "the smaller, submerged communities" because they "have to offset their lack of a long, rich, continuous ethno-history through 'cultural wars' in which philology, archaeology, anthropology, and other scientific disciplines are used to trace uncertain genealogies, to root populations in their native terrains, to document their distinctive traits and cultures, and to annex earlier civilizations." A. Smith, *National Identity*, 165.

11. Plakans, *The Latvians*, 94. The National Awakening thus represents Phase B in Hroch's periodization of nation formation: the phase of "patriotic agitation" by the intelligentsia (the New Latvians) which follows the period of limited "scholarly interest" and precedes the "rise of a mass national movement."

12. Erich E. Haberer, "Economic Modernization and Nationality in the Russian Baltic Provinces 1850–1900," *Canadian Review of Studies in Nationalism* 12, no. 1 (1985): 166.

13. Plakans, *The Latvians*, 97. See also Plakans, "The Latvian National Awakening," *Bulletin of Baltic Studies* 1, no. 6 (1971).

14. Ivo Banac and Katherine Verdery, eds., *National Character and National Ideology in Interwar Eastern Europe* (New Haven: Yale Center for International and Area Studies, 1995).

15. Tamas Hofer, "The Creation of Ethnic Symbols from the Elements of Peasant Culture," in *Ethnic Diversity and Conflict in Eastern Europe*, ed. Peter F. Sugar (Oxford and Santa Barbara: ABC-Clio, 1980).

16. Frances Millard, "Nationalism in Poland," in *Contemporary Nationalism in East Europe*, ed. Paul Latawski (New York: St. Martin's Press, 1995), 108.

17. Keely Stauter-Halsted, *The Nation in the Village: the Genesis of Peasant National Identity in Austrian Poland, 1848–1914* (Ithaca: Cornell University Press, 2001), 107.

18. Hofer, *Creation of Ethnic Symbols*, 119.

19. Jerzy Jedlicki, "Polish Concepts of Native Culture," in Banac and Verdery, *National Character*, 4.

20. Katherine Verdery, "National Ideology and National Character in Interwar Romania," in Banac and Verdery, *National Character*, 111. See also Verdery, *National Ideology Under Socialism: Identity and Cultural Politics in Ceausescu's Romania* (Berkeley and Los Angeles: University of California Press, 1991).

21. Jedlicki, "Polish Concepts of Native Culture," 4–5.

22. Iver Neumann, *Russia and the Idea of Europe* (London: Routledge, 1996).

23. Cathy A. Frierson, *Peasant Icons: Representations of Rural People in Late Nineteenth Century Russia* (Oxford: Oxford University Press, 1993), 182.

24. Larry Wolff, *Inventing Eastern Europe: The Map of Civilization on the Mind of the Enlightenment* (Stanford: Stanford University Press, 1994), 9, 13.

25. Mark Bassin, "Russian Geographers and the 'National Mission' in the Far East," in *Geography and National Identity*, ed. David Hooson (Oxford: Blackwell, 1994), 129.

26. O. Čakars, A. Grigulis, M. Losberga, *Latviešu literatūras vēsture no pirmsākumiem līdz XIX gadsimta 80. gadiem* (Riga: Zvaigzne, 1990), 253.

27. Douglas R. Weiner, *Models of Nature: Ecology, Conservation and Cultural Revolution in Soviet Russia* (Bloomington: Indiana University Press, 1988).

28. Barnard, *Herder's Social and Political Thought*, 70.

29. The most prominent articulators of *völkisch* environmental determinism were Alexander von Humboldt, Wilhelm Heinrich Riehl, and Friedrich Ratzel. See Raymond H. Dominick III, *The Environmental Movement in Germany: Prophets & Pioneers, 1871–1971* (Bloomington: Indiana University Press, 1992), 22–23.

30. John Alexander Williams, "The Chords of the German Soul Are Tuned to Nature," *Central European History* 29, no. 3 (1996): 345–47. See also Celia Applegate, *A Nation of Provincials: The German Idea of Heimat* (Berkeley and Los Angeles: University of California Press, 1990).

31. Ultimately, of course, German conservationism was co-opted by the Nazi regime, and environmental determinism gave way to full-fledged racialism. In the late Weimar years, conservationists began to celebrate the natural Heimat as "clean" and "healthy" and as an antidote to the frightening chaos of mass urban society and a restorer of the war-torn social fabric. By the 1930s, Nazis began "attempting to eradicate 'Jewish influence' in the natural landscape itself." Williams, "The Chords of the German Soul," 353, 381. See also, for example, Janet Biehl and Peter Staudenmaier, *Ecofascism: Lessons from the German Experience* (Edinburgh, San Francisco: AK Press, 1995); Anna Bramwell, *Ecology in the 20th Century* (New Haven: Yale University Press, 1989); and Paul Josephson, *Totalitarian Science & Technology* (Atlantic Highlands, NJ: Humanities Press, 1996).

32. Anne Buttimer, "Edgar Kant and Balto-Skandia: *Heimatkunde* and Regional Identity," in Hooson, *Geography and National Identity*, 162. In an interesting twentieth-century parallel, *Heimatkunde* was central to the Zionist project in Palestine and "became the pivot around which the entire Israeli educational system would revolve." Known as *yidi'at ha-arets* (knowing the land), the teaching of natural history and geography became "much more than just a school subject; a host of overlapping institutions were in large part dedicated to its study: research institutes, the Society for the Preservation of Nature in Israel, youth movements, paramilitary units, the army, and, not least, the literary establishment." Todd Hasak-Lowy, "Thesis, Antithesis, Thesis: Nature in S. Yizhar's War of Independence Stories," in *Crisis and Memory: The Representation of Space in Modern Levantine Narrative*, ed. Ken Seigneurie (Wiesbaden: Reichert Verlag, 2003), 133.

33. O. Kronwald, *Der Unterricht in der Heimatskunde* (Dorpat, 1867). A Latvian translation by Bebru Juris was published in 1922 in Riga under the title *Dzimtenes Mācība*.

34. Atis Kronvalds, *Tagadnei: Izlase* (Riga: Liesma, 1987), 56.

35. Ibid., 57.

36. Jāzeps Rudzītis, introduction to Kronvalds, *Tagadnei*, 3.

37. Ibid., 7.

38. "Mežs un koki" (1868), in Kronvalds, *Tagadnei*, 246.

39. "Rudens nāk, rudens nāk!" (1869), in Kronvalds, *Tagadnei*, 249.

40. Ibid., 247.

41. Čakars et al, *Latviešu literatūras vēsture*, 224.

42. Plakans, "The Latvians," in *Russification in the Baltic Provinces and Finland, 1855–1914*, ed. Edward C. Thaden (Princeton: Princeton University Press, 1981), 221.

43. Paulis Lazda, "The Phenomenon of Russophilism in the Development of Latvian Nationalism in the Nineteenth Century," in *National Movements in the Baltic Countries during the Nineteenth Century*, ed. Aleksander Loit (Stockholm: Seventh Conference on Baltic Studies in Scandinavia, 1983), 129. Lazda reports that Kaspars Biezbardis, the first Latvian graduate of Tartu University and among the first generation of New Latvians, believed Latvians were ethnically close relatives of the Slavs—"half-brothers" or even "*lettische Slaven*"—and "advocated the 'return' of the Latvian language to the Cyrillic alphabet" (130).

44. Ibid., 131.

45. Krišjānis Valdemārs, *Tevzemei* (Riga: Avots, 1991), 80.

46. Ibid., 106.

47. Ibid., 118. Valdemārs thus serves as a counterexample to David Laitin's hypothesis that bilingual cosmopolitan nationalists seek to keep peasants in the periphery monolingual, so that the former can continue to profit as linguistic "monopoly mediators." Laitin, *Identity in Formation*, 338.

48. Valdemārs, *Tevzemei*, 22.

49. Quoted in Kārlis Lūsis and Gints Šimanis, eds., *Krišjānim Valdemāram 170. Apceres, raksti, vēstules* (Riga: Latvijas Jūrniecības fonds, 1995), 18.

50. Ibid., 21.

51. Ibid.

52. Quoted in ibid., 56.

53. Lūsis and Šimanis, *Krišjānim Valdemāram 170*, 23. The commercial significance of maritime trade helps explain the great interest aroused within Russian government circles by Valdemārs's seafaring advocacy: in 1851 88% of Russia's total exports passed through seaports, and 59% through ports on the Baltic. Arveds Švābe, *Latvijas vēsture 1800–1914* (Uppsala: Daugava, 1962), 378.

54. Valdemārs, "Jūrniecības kopšana Baltijā," in Lūsis and Šimanis, *Krišjānim Valdemāram 170*, 36–39.

55. Lūsis and Šimanis, *Krišjānim Valdemāram 170*, 119. Not all of the New Latvians shared Valdemārs's optimism in this regard. Atis Kronvalds, for example, "was more aware of the dangers of a loss or dilution of the language and culture if increased social mobility was not accompanied by an effective elementary education in Latvian." David Kirby, *The Baltic World 1772–1993: Europe's Northern Periphery in an Age of Change* (Longman: London and New York, 1995), 128.

56. A. Upīts, quoted in Ausma Cimdiņa, "'Sava kaktiņa, sava stūrīša zemes' sakāma vārda tiesības, jeb kāpēc šo stāstu neizlasīt līdz galam?" *Karogs* 12 (1998): 121.

57. Matīss Kaudzīte, quoted in ibid., 119.

58. Jānis Purapuķe, *Savs kaktiņš—savs stūrītis zemes* (Lübeck: J. Sins, 1948), 35.

59. Plakans, "Peasants, Intellectuals, and Nationalism in the Russian Baltic Provinces, 1820–90," *Journal of Modern History* 46 (September 1974): 462.

60. Plakans, *The Latvians*, 103.

61. Bruno Kalnins, "The Social Democratic Movement in Latvia," in *Revolution and Politics in Russia: Essays in Memory of B. I. Nicolaevsky*, ed. Alexander and Janet Rabinowitch, with Ladis K. D. Kristof (Bloomington: Indiana University Press, 1972), 137.

62. Kirby, *The Baltic World*, 232, 237.

63. By comparison, the figures were 45% in Petrograd and 24% in Russia overall. Georg von Rauch, *The Baltic States: The Years of Independence. Estonia, Latvia, Lithuania 1917–1940*, 2nd ed. (New York: St. Martin's Press, 1995), 36.

64. On the war years in Latvia, in addition to the above-cited works by Švābe, Plakans, Rauch, and Kirby, see also Andrievs Ezergailis, *The Latvian Impact on the Bolshevik Revolution* (Boulder: East European Monographs, 1983) and *The 1917 Revolution in Latvia* (Boulder: East European Monographs, 1974).

65. Ghita Ionescu, "Eastern Europe," in *Populism: Its Meaning and National Characteristics*, ed. Ghita Ionescu and Ernest Gellner (New York: Macmillan, 1969), 97, 99.

66. The Social Democrats dropped from 39% of the popular vote in 1920 to 21% in 1931, as compared to 17% and 14%, respectively, for the Farmers' Union, while ethnic minority parties' share grew from 11% to 17%. Adolfs Šilde, *Latvijas vēsture 1914–1940* (Stockholm: Daugava, 1976), 394. In 1920 ethnic Latvians comprised 72.7% of the population; the largest minorities were Russians (7.8%), Germans (3.6%), and Jews (5%). Plakans, *The Latvians*, 132.

67. Kirby, *The Baltic World*, 293.

68. Ronald Grigor Suny argues compellingly for historical contingency in explaining why in 1918 Latvia became a nation-state and not a Soviet Socialist Republic, in *The Revenge of the Past: Nationalism, Revolution, and the Collapse of the Soviet Union* (Stanford: Stanford University Press, 1993), chap. 2. However, the culturally nationalist orientation of even Latvian socialists in the prewar period, as well as the politically nationalist stance of other agitators, belies Hobsbawm's overdrawn claim: "The victorious Germans set up three small Baltic nation-states for which there was no historical precedent at all, and— at least in Estonia and Latvia—no noticeable national demand." Hobsbawm, *Nations and Nationalism since 1780*, 165.

69. Aldis Purs, "Latvians as an Imagined Community," presented at the annual meeting of the Association for the Study of Nationalities, Columbia University, 1998, 16.

70. Paraphrased in Artūrs Boruks, *Zemnieks, zeme un zemkopība Latvijā: no senākiem laikiem lidz mūsdienām* (Riga: Grāmatvedis, 1995), 246.

71. Quoted in Šilde, *Latvijas vēsture*, 370.

72. Ibid.

73. Arnolds Aizsilnieks, *Latvijas saimniecības vēsture 1914–1945* (Sweden: Daugava, 1968), 238.

74. Heinrihs Strods, *Latvijas lauksaimniecības vēsture: no vissenākajiem laikiem līdz XX gs. 90. gadiem* (Riga: Zvaigzne, 1992), 160, 174.

75. Markus Gailītis, quoted in Šilde, *Latvijas vēsture*, 368.

76. Stukuls Eglitis, *Imagining the Nation*, 185.

77. Purs notes that despite the popular understanding of the *viensēta* as a single-family homestead, in fact many farmhands and other unrelated individuals also lived and worked in the typical homestead (personal communication).

78. Purs, "Latvians as an Imagined Community," 19–20.

79. Purs notes that "the 'father' of the museum, Pauls Kundziņš, would later take this idea to the extreme, writing in exile (post-WWII) that the history of the Latvian homestead is the history of Latvia" (ibid., 20). See Pauls Kundziņš, *Latvju sēta* (Sweden: Daugava, 1974).

80. Latvian Academy of Sciences Institute of Literature, Folklore, and Art, *Latviešu literatūras vēsture. 2. sējums: 1918–1945* (Riga: Zvaigzne ABC, 1999), 87, 113, 239–40, 330–33.

81. A. Dravnieks, *Latviešu literaturas vēsture* (Goppingen, Germany: Gramatu draugs, 1946), 113.

82. Jānis Jaunsudrabiņš, *Jaunsaimnieks un velns* (Stockholm: Daugava, 1949), 104.

83. Ibid., 18.

84. Ibid., 89.

85. Latvian Academy of Sciences, *Latviešu literatūras vēsture*, 90.

86. Jānis Jaunsudrabiņš, *Baltā grāmata* (Riga: Liesma, 1971), 27.

87. Garlībs Merķelis, *Izlase* (Riga: Liesma, 1969), 61.

88. Jānis Akurāters, *Kalpa zēna vasara: atmiņas zīmējums* (Riga: Latvijas valsts izdevniecība, 1956), 25, 65, 73.

89. Dravnieks, *Latviešu literaturas vēsture*, 174. The merging of nature with national authenticity is reinscribed by the literary historian Dravnieks, who published his *History of Latvian Literature* while a refugee in the displaced persons camps of Germany. Of the great lyric poet and "king of the Latvian fairy tale," Kārlis Skalbe, for example, Dravnieks writes: "The Latvian land, nature, life, and the destiny of the *tauta* are the eternal springs which give Skalbe's poetry its Latvian content and pleasant expression. . . . There is noth-

ing artificial or foreign [in Skalbe's poetry], it has the natural beauty, unique fragrance and vigorous life of simple country flowers" (186).

90. Agrarian nationalism was not the only orientation of Latvian writers expressing closeness to nature. Poets in particular often conceived of a transcendental merging of self with nature, unmediated by labor. For example, the poetry of Vilis Plūdons is "characterized by a deep, pantheistic sense of nature, a merging with nature." Kārlis Skalbe "does not worship nature, and thus does not oppose himself to nature, as would be characteristic of a Romantic, but feels connected to it through a blood relationship." For Kārlis Jekabsons, "merging with nature and feelings of love allow him to sense the presence of eternity in everything temporal and impermanent." Latvian Academy of Sciences, *Latviešu literaturas vēsture*, 151, 204–05.

91. Rūdolfs Blaumanis, *Indrāni* (Rockville, MD: ALA Kultūras birojs, 1965), 66.

92. K. Birnbaums, "Meža dienas Latvijā 1934. gadā," *Mežsaimniecības rakstu krājums XII* (Riga: Latvijas Mežkopju savienība, 1934), 166.

93. K. Birnbaums, "Rīkosim meža dienas!" *Meža Dzīve* 1936, 12: 4539.

94. Kārlis Ulmanis, "Šis darbs apstiprinās mūsu labo gribu un nedziestošo sirds kvēli," *Meža Dzīve* 1936, 129: 4587.

95. J. Ozols, "Meža dienas Latvijā 1939. gadā," *Mežsaimniecības rakstu krājums XVII* (Riga: Latvijas Mežkopju savienība, 1939), 205.

96. Aizsilnieks, *Latvijas saimniecības vēsture*, 248.

97. Ibid., 349–61.

98. Quoted in Kirby, *The Baltic World*, 293.

99. J. Ruņģis, paraphrased in Boruks, *Zemnieks, zeme un zemkopība*, 251.

100. Quoted in ibid., 251–52.

101. Aizsilnieks, *Latvijas saimniecības vēsture*, 272, 422, 586.

102. Miķelis Valters, quoted in Edgars Dunsdorfs, *Kārļa Ulmaņa dzīve: Ceļinieks, politiķis, diktators, moceklis* (Sweden: Daugava, 1978), 36.

103. Plakans, *The Latvians*, 134.

104. Kirby, *The Baltic World*, 326. Fascist racialism and extreme xenophobia in interwar Latvia were largely urban phenomena, represented by the Pērkoņkrusts (Thunder Cross) movement, established in 1931 on the basis of the Latvian National Club (est. 1922), under the slogan "Latvia for the Latvians." Like the Nazis, Pērkoņkrusts celebrated romantic notions of the simple pastoral life. However, unlike the Ulmanis cohort, whose main base of support was the countryside, support for Pērkoņkrusts, "despite its members' great love for things rural . . . was restricted primarily to the urban areas of the country." Rauch, *The Baltic States*, 152.

105. Aizsilnieks, *Latvijas saimniecības vēsture*, 711.

106. Between 1934 and 1939, 290 kilometers of rivers were regulated and 9.3 million lats were spent on regulating rivers and digging drainage ditches, of which 80% was paid by the state. Ibid., 352, 724.

107. Jaunsudrabiņš, *Jaunsaimnieks un velns*, 101. Interestingly, in Mārtiņš Zīverts's 1931 play *Oil*, a rare instance in interwar literature of an ecological orientation to nature, the entirely peaceable and good-natured Swamp Devil is portrayed as a steward of nature. It is the Swamp Devil who voices the ecological critique of man's rapacious exploitation of natural harmony: "You have waged war against nature, in order to rape it and make it your slave. You want to create a paradise on earth, but it's all lies, lies, lies! You have created hell." Latvian Academy of Sciences *Latviešu literatūras vēsture*, 347. Jaunsudrabiņš's swamp devil is also the defender of unspoiled nature, but viewed through the agrarian lens, this defense becomes an act of selfishness and destruction. A utilitarian attitude toward wetlands is not incompatible with the *völkisch* reverence for nature, as Atis Kronvalds's remark suggests: "When the young generation have become acquainted with the pleasant aspects of their fatherland, they should then be informed about the inadequacies of nature in our country, so that the young generation can learn to acknowledge that we must be humble and that we still have many swamps to destroy, before a fertile foundation is laid everywhere; thus the young generation will learn to evaluate what still remains for us, as the inheritors of the fatherland, to do." Kronvalds, *Tagadnei*, 57.

108. Aizsilnieks, *Latvijas saimniecības vēsture*, 724.

109. Overt celebration of the Ulmanis regime was, of course, unacceptable in Latvia during the Soviet period, but it flourished among Latvians in exile. For example, the émigré historian Adolfs Šilde, whose career began in interwar Latvia, was critical of the authoritarian Ulmanis regime overall, but he claimed approvingly that "the raising of the peasant class to a place of honor found widespread support in society" and "raised Latvians' self-esteem." About the agrarian reform he wrote: "It was not only economic interests and the desire to obtain property that emerged into the light of day during the agrarian reform, but an almost metaphysical and sacred attachment to the land, which had given and promised to continue to give the tender of the land happiness and satisfaction in his work." Šilde dismissed interwar literary portrayals of the darker aspects of rural life as products of false consciousness that, in focusing on "the grueling daily life and hardships of the *jaunsaimnieks*, concealed the irrepressible joy that seized the new farmers in their constructive labor and the cultivation of 'their little corner, their little piece.'" Šilde, *Latvijas vēsture*, 689, 370, 523.

110. Ibid., 736.

111. Ibid., 833, 347. One contemporary analyst calculated that farms of 2–5 hectares employed 6.3 times as many people per hectare as farms of 50–100 hectares. Boruks, *Zemnieks, zeme un zemkopība*, 303.

112. Aizsilnieks, *Latvijas saimniecības vēsture*, 251. Timber accounted for 30–36% of exports throughout the period. Aizsilnieks concurs with some contemporary critics that it would have been more efficient to use timber export revenues to finance the manufacture of fire-resistant building materials.

113. Ibid., 725. Average grain yield was less than half the level of the most efficient producers, such as Denmark. Brian Van Arkadie and Mats Karlsson, *Economic Survey of the Baltic States* (New York: New York University Press, 1992), 63.

114. Total and per capita GNP in 1937 were 94% and 93%, respectively, of the 1934 figures, rebounding only to 103% and 101% in 1938. Industrial productivity also declined between 1934 and 1938. Aizsilnieks, *Latvijas saimniecības vēsture*, 834, 757.

115. A. Zušēvics, quoted in Boruks, *Zemnieks, zeme un zemkopība*, 248.

116. Quoted in Aizsilnieks, *Latvijas saimniecības vēsture*, 721.

117. Quoted in ibid., 720.

118. In 1935 Ulmanis's minister of agriculture proposed requiring mandatory agricultural labor service from the entire population, but deep hostility to this suggestion led the government first to try to recruit laborers through economic incentives such as family support payments and subsidized loans for housing construction. In May 1939 a new law imposed residency requirements on applicants for all urban jobs. In 1940 the government further mandated that all students without full-time employment were to spend two months of the summer in practical, preferably agricultural, labor, and announced that any unemployed or underemployed persons between the ages of sixteen and sixty, as well as noncitizens, could be drafted into economic service for up to six months. Aizsilnieks, *Latvijas saimniecības vēsture*, 733–34, 810.

119. Ibid., 810.

120. While the GNP share of agriculture, forestry, and fisheries diminished from 72% in 1933 to 66% in 1941, that of industry and the trades grew from 28% to 34%. And if the contribution of Latvia's state-owned forestry sector is subtracted, then agriculture's share of GNP barely exceeded that of industry. Strods, *Latvijas lauksaimniecības vēsture*, 162.

Chapter 2 | "The Occupation of Beauty"

1. Strods, *Latvijas Lauksaimniecības vēsture*, 181.

2. On Stalin's class war against the peasantry, see, for example, Robert Conquest, *The Harvest of Sorrow: Soviet Collectivization and the Terror-Famine* (New York: Oxford University Press, 1986); Sheila Fitzpatrick, *Stalin's Peasants: Resistance and Survival in the Russian Village after Collectivization* (New York: Oxford University Press, 1994); and Moshe Lewin, *Russian Peasants and Soviet Power: A Study of Collectivization* (Evanston: Northwestern University Press, 1968).

3. In Latvia, a kulak was any farmer who employed or had ever employed permanent salaried labor; systematically employed or had employed seasonal or day labor; earned or had ever earned income from a mill, dairy, or other enterprise powered by machines, water, or steam, or from the rent thereof; earned or had ever earned income from the purchase and resale of goods or from usury; or had supported the German occupation. Boruks, Zemnieks, zeme un zemkopība, 328.

4. Romuald J. Misiunas and Rein Taagepera, The Baltic States: Years of Dependence, 1940–1990, 2nd ed. (Berkeley and Los Angeles: University of California Press, 1993), 83–94.

5. Strods, Latvijas Lauksaimniecības vēsture, 223–24.

6. Kirby, The Baltic World, 412.

7. Strods, Latvijas Lauksaimniecības vēsture, 248.

8. Ibid., 250.

9. Misiunas and Taagepera, The Baltic States, 112.

10. Strods, Latvijas Lauksaimniecības vēsture, 247; Boruks, Zemnieks, zeme un zemkopība, 363.

11. According to Soviet statistics, the shares of net material product of agriculture and industry were 12% and 56%, respectively, in 1980, and 24% and 45% in 1989. The employment shares underwent a full reversal from 66% for agriculture and 13% for industry in 1930, to 16% and 40%, respectively, in 1987. Van Arkadie and Karlsson, Economic Survey of the Baltic States, 70. (The authors note, however, that Soviet statistics may not be accurate or meaningful.)

12. Strods, Latvijas Lauksaimniecības vēsture, 66.

13. Ibid., 169.

14. Boruks, Zemnieks, zeme un zemkopība, 361. On the Soviet land reclamation program and other reforms of the 1960s, see Thane Gustafson, Reform in Soviet Politics: Lessons of Recent Policies on Land and Water (Cambridge: Cambridge University Press, 1981); Zhores Medvedev, Soviet Agriculture (New York: Norton, 1987); and Alec Nove, Soviet Agriculture: The Brezhnev Legacy and Gorbachev's Cure (Los Angeles: Rand/UCLA Center for the Study of Soviet International Behavior, 1988).

15. See Robert G. Darst Jr., "Environmentalism in the USSR: The Opposition to the River Diversion Projects," Soviet Economy 4, no. 3 (1988): 223–52; Joan DeBardeleben, The Environment and Marxism-Leninism (Boulder: Westview Press, 1985); Murray Feshbach and Alfred Friendly Jr., Ecocide in the USSR: Health and Nature under Siege (New York: Basic Books, 1992); Loren Graham, Science in Russia and the Soviet Union: A Short History (Cambridge: Cambridge University Press, 1993); Gustafson, Reform in Soviet Politics; David Joravsky, The Lysenko Affair (Cambridge: Harvard University Press, 1970); Josephson, Totalitarian Science & Technology; Boris Komarov, The Destruction of Nature in the Soviet Union (White Plains, NY: M. E. Sharpe, 1980); D. J. Peterson, Troubled Lands: The Legacy of Soviet Environmental Destruction (Boulder: Westview Press, 1993); Weiner, Models of Nature; Weiner, A Lit-

tle Corner of Freedom: Russian Nature Protection from Stalin to Gorbachev (Berkeley and Los Angeles: University of California Press, 1999); and Charles E. Ziegler, *Environmental Policy in the USSR* (Amherst: University of Massachusetts Press, 1987).

16. Juris Dreifelds, "Latvia," in *Environmental Resources and Constraints in the Former Soviet Union*, ed. Philip R. Pryde (Boulder: Westview Press, 1995), 116.

17. Strods, *Latvijas Lauksaimniecības vēsture*, 256.

18. J. Rutkis, ed., *Latvia: Country and People* (Stockholm, 1967); quoted in Aija Melluma, "Metamorphoses of Latvian Landscapes during Fifty Years of Soviet Rule," *Geojournal* 33, no. 1 (1994): 56.

19. Vilis Lācis, *Uz jauno krastu* (Riga: Latvijas valsts izdevniecība, 1958). A Communist Party member since 1928 and "formerly a prosperous petty-bourgeois author of adventure novels," Lācis served as chairman of the Latvian SSR Council of Ministers in the 1940s and 1950s. Rolfs Ekmanis, *Latvian Literature Under the Soviets, 1940–1975* (Belmont, MA: Nordland, 1978), 57.

20. A. Pope, "Ar meliorāciju saistitie dabas aizsardzības jautājumi," *Dabas un vestures kalendārs* (DVK) 1973, 127–31.

21. G. Sēne, "Lubāna pārvērtības," DVK 1978, 153; see also Mirdza Leinerte, "Lubāns—pagātne, tagadne, nākotne: gandrīz teika," DVK 1998, 191–94.

22. Melluma, "Metamorphoses of Latvian Landscapes," 56, 61.

23. Strods, *Latvijas Lauksaimniecības vēsture*, 240.

24. Ibid., 245.

25. Ibid., 242.

26. Boruks, *Zemnieks, zeme un zemkopība*, 361.

27. V. Ozola, "Jauno kultūras ainavu veidojot," DVK 1976, 190. See also I. Dulbe, "Meliorācija un ainava," DVK 1973: 124–27; and V. Ozola, "Laikmets un ainava," DVK 1974, 121–23.

28. R. Tavare, "Saudzīga attieksme pret dabu latviešu folkloras skatījumā," DVK 1977, 88–93. See also "Mana tautasdziesma uzrunā astronomu, dravnieku [etc.]," DVK 1985, 142–53; and Valdemārs Ancītis, "Meža māte," DVK 1987, 119–21.

29. Valentin Rasputin, "What We Have: A Baikal Prologue without an Epilogue," in *Siberia on Fire* (DeKalb: Northern Illinois University Press, 1989), 196.

30. Thomas B. Rainey, "Siberian Writers and the Struggle to Save Lake Baikal," *Environmental History Review* 15, no. 1 (1991): 48–60.

31. Quoted in Weiner, *A Little Corner of Freedom*, 335.

32. For example, L. Suna, "Pilsētu zaļo zonu veidošanas principi," *Jaunākais mežsaimniecībā* 1970, no. 12: 57.

33. Dulbe, "Meliorācija un ainava," 125–26.

34. Anon., "Koki—dabas pieminekļi Latvijā," *Meža Dzīve*, 1938, no. 4: 179–80. In pre-Soviet Estonia, closeness to nature was similarly seen as central to national identity, and glacial erratic boulders figured prominently, like great trees in Latvia, as markers of that identity. Rob Smurr, "Nationalizing Nature: the history, preservation and meaning of glacial erratic boulders in Estonia," presented at the annual meeting of the American Association for the Advancement of Slavic Studies, Pittsburgh, November 2002.

35. I have also seen *dižkoks* translated as "noble tree"; other alternatives include "ancient" and "giant," although neither of these captures the full range of meanings of the Latvian adjective *dižs*. The Russian equivalent is *vekovoe derevo*, or "ancient tree."

36. Staņislāvs Saliņš, "Saudzēsim vecos dižkokus," *Jaunākais mežsaimniecībā*, 1962, no. 3: 68. The following minimum diameters were required to qualify as a *dižkoks* by size: oak, 4 meters; pine, 3.5; ash, fir, aspen, linden, willow, 3; black alder, elm, maple, birch, 2.5; juniper, 0.8; yew tree, any size. Saliņš, "Latvijas PSR dižkoki un retie koki," *Jaunākais mežsaimniecībā*, 1970, no. 12: 69.

37. Guntis Eniņš, "Latvijas dižkoku iznīcināšanas vēsture," *Meža Dzīve*, 1975, no. 1: 33.

38. Saliņš, *Latvijas dižkoki un retie koki* (Riga: Zinātne, 1974).

39. Eniņš, *Koks—Dabas piemineklis* (Riga: Zinātne, 1982), 17.

40. Eniņš, "Dižkoki Baltijas republikās," *Mežsaimniecība un mežrūpniecība*, 1978, no. 1: 26–30.

41. Eniņš, "Slinkums? Problēma? Nezināšana?" *Padomju Jaunatne*, February 19, 1978; Eniņš, interview by the author, Riga, July 13, 1999.

42. Zigmunds Skujins, "Imants Ziedonis Opens Clocks," trans. Juris Silenieks, *World Literature Today* 72, no. 2 (1998): 298.

43. Māra Zālīte in Imants Ziedonis, *Tutepatās. Kopotie raksti* (Riga: Nordik, 1998), no. 9: 385. A tremendously popular poet, playwright, and librettist herself, Zālīte also became one of the leading voices of the independence movement.

44. Skujins, "Imants Ziedonis Opens Clocks," 297.

45. Ziedonis, *Kopotie raksti*. Ziedonis compiled the book in the late 1980s from journal entries; it was republished in 1998 as volume 9 of his collected writings.

46. Eniņš, *Koks—Dabas piemineklis*, 24.

47. Janīna Kursīte, *Mītiskais folklorā, literatūrā, mākslā* (Riga: Zinātne, 1999), 182, 201–02. See also Vilius Audrius Dundzila, "Baltic Indigenous Religion," in *Encyclopedia of Religion and Nature*, ed. Bron Raymond Taylor (London: Continuum, 2005).

48. Eniņš, "Ozolīti, zemzarīti," *Oktobra karogs*, May 15, 1979, 3.

49. Eniņš, "Latvija ir un būs dižkoku zeme," *Lauku Avīze*, September 6, 1994, 28.

50. Eniņš, *Koks—Dabas piemineklis*, 28.

51. Ziedonis, *Kopotie raksti*, 49.

52. Eniņš, 1999 interview.

53. Eniņš, "Latvija ir un būs."

54. Ziedonis, *Kopotie raksti*, 305.

55. Eniņš, "Ozoli mums blakus," *Lauku Dzīve*, 1982, 19.

56. Eniņš, "Tēvu sētas dārgumi tavās paša rokās," DVK 1994, 86. Saliņš, "Saudzēsim vecos dižkokus," 70. Saliņš, "Latvijas PSR dižkoki," 58.

57. Ziedonis, *Kopotie raksti*, 25.

58. Ibid., 290.

59. Ibid., 49–50.

60. Rasputin, *Siberia on Fire*, 109–10.

61. Ibid., 177.

62. "It was not blood but language which Herder regarded as the essential criterion of a *Volk*." Barnard, *Herder's Social and Political Thought*, 70.

63. Ziedonis, *Kopotie raksti*, 231.

64. Mārtins Dāboliņš, "Par dižkoku, reto un savdabīgo koku uzskaiti Latvijas PSR," *Mežsaimniecība un mežrūpniecība*, 1980, no. 5: 15–18.

65. Ziedonis, *Kopotie raksti*, 263.

66. Ibid., 53.

67. Eniņš, *Koks—Dabas piemineklis*, 66. Ziedonis, *Kopotie raksti*, 49.

68. Eniņš, "Slinkums?" Eniņš, *Tepat Latvijā* (Riga: Liesma, 1984), 26.

69. Eniņš, "Ozolīti, zemzarīti."

70. Epigraph to Saliņš, *Latvijas dižkoki*.

71. Ziedonis, *Kopotie raksti*, 11.

72. Ibid., 212, 197, 356. Ziedonis, interview by the author, Riga, August 9, 1999.

73. Dainis Īvāns and Artūrs Snips, "Par Daugavas likteni domājot," *Literatūra un Māksla*, October 17, 1986.

74. Nils R. Muiznieks, "The Daugavpils Hydro Station and *Glasnost* in Latvia," *Journal of Baltic Studies* 18 (Spring 1987): 667. See also Marshall I. Goldman, "Environmentalism and Nationalism: An Unlikely Twist in an Unlikely Direction," in *The Soviet Environment: Problems, Policies and Politics*, ed. John Massey Stewart (Cambridge: Cambridge University

Press, 1992); Barbara Jancar, "The Environmental Attractor in the Former USSR: Ecology and Regional Change," in *The State and Social Power in Global Environmental Politics*, ed. Ronnie D. Lipschutz and Ken Conca (New York: Columbia University Press, 1993); Pryde, "The Environmental Basis for Ethnic Unrest in the Baltic Republics," in Massey Stewart, *The Soviet Environment*; and Charles E. Ziegler, "Political Participation, Nationalism and Environmental Politics in the USSR," in Massey Stewart, *The Soviet Environment*.

75. Jane I. Dawson, *Eco-Nationalism: Anti-nuclear Activism and National Identity in Russia, Lithuania, and Ukraine* (Durham: Duke University Press, 1996), 29. Not all environmental protest movements of the late Soviet period followed this pattern. In Kazakstan, for example, the movement to halt nuclear testing at Semipalatinsk assumed, instead, what an Edward A. D. Schatz calls an "eco-internationalist" form. Movement leaders eschewed anticolonial rhetoric, emphasizing instead the global dimensions of the problem; they traveled and formed links to similar groups abroad, hosted international conferences and secured international funding. Schatz argues that this internationalist strategy was more popularly resonant in Kazakstan because of the particular cultural and socioeconomic legacies of Soviet rule there. Edward A. D. Schatz, "Notes on the 'Dog That Didn't Bark': Eco-Internationalism in Late Soviet Kazakstan," *Ethnic and Racial Studies* 22, no. 1 (1999): 156.

76. Kirby, *The Baltic World*, 430–31; see also Pryde, "Environmental Basis for Ethnic Unrest." The Daugavpils dam protest is discussed in tandem with the campaigns against the Ignalina nuclear plant in Lithuania and phosphorite mining in Estonia.

77. Beissinger, *Nationalist Mobilization*, 32.

78. Dawson, *Eco-Nationalism*, 162.

79. Goldman, "Environmentalism and Nationalism." My interpretation of the Latvian case thus supports Andrew Tickle and Ian Welsh's claim that "the environment is more strongly bound into the structures and values of these [central and east European Communist] societies than the surrogacy theory would suggest." "Environmental Politics, Civil Society and post-Communism," in Tickle and Welsh, *Environment and Society*, 159.

80. Viktors Avotiņš, "Domu Daugava. Sirdsdaugava," *Literatūra un Māksla*, November 21, 1986.

81. All of the antidam articles that appeared in the Latvian press at the time, as well as several not published due to censorship, were compiled by Dainis Īvāns and Artūrs Snips and later published as *Domu Daugava, Sirdsdaugava* (Riga: Avots, 1989).

82. Juris Dreifelds, "Two Latvian Dams: Two Confrontations," *Baltic Forum* 6, no. 1 (1989): 13. Janīna Kursīte observes that in Latvian belletristic literature as well, "the centuries-old archetypal Daugava image was preserved in the consciousness, and even more the unconscious of the Latvian reader. Every now and then (and especially during the period of national awakening) this image was revived, through overly direct reminders of the sacrality of the Daugava." Kursīte *Mītiskais folklorā*, 104.

83. Vera Kacēna, "Nedrīkstam pārsteigties," *Literatūra un Māksla*, March 22, 1958.

84. Dreifelds, "Two Latvian Dams," 13.

85. Īvāns and Snips, "Par Daugavas likteni."

86. Ibid.

87. Artūrs Mauriņš, quoted in Zigfrīds Dzedulis, "Zaļā suņa alternatīva," *Literatūra un Māksla*, November 14, 1986.

88. Guntis Eberhards, "Daugavas programma," *Latvijas Kultūras fonda gaduraksti, 1987–1988* (Riga: Avots, 1991), 39.

89. Kursīte, *Mītiskais folklorā*, 378 n.

90. *Dabas retumu krātuve*, brochure.

91. Eniņš, letter to the editor, *Diena*, October 6, 1995.

92. Dzintars Driba, "Harmonija latviešu tautas arhitektūrā," *DVK* 1992, 135–36.

93. Ilze Janele, "Ar skatu—uz savu kultūrainavu," *DVK* 1990, 217–20. Andris Salna, "Latviešu lauku sētas psihokomforts," *DVK* 1990, 220. Driba, "Harmonija latviešu tautas arhitektūrā," 135–36.

94. Edgars Mucenieks, "Mūsu dzīves pamatu dzīves ziņa," *DVK* 1992, 132–33.

95. Aldis Pūtelis, "Pasaule ap mums un tās atspulgs 'Latvju dainās'," *DVK* 1994, 70–75.

96. Valdemārs Ancītis, "Latvieša svētvietas un svētumi," *DVK* 1993, 219–24.

97. Beitnere, "Vai latvieši ir tikai zemnieki?" 159.

98. Ārija Elksne, "Lai sauktos par latvieti," *DVK* 1992, 143.

99. Ārija Karpova, "Tēva mājas un latvieša raksturs," *DVK* 1993, 214–19. Jēkabs Raipulis, "Latviešu labās un sliktās īpašības," *DVK* 1993, 205–14. Janele, "Ar skatu." Driba, "Harmonija latviešu tautas arhitektūrā." Ārija Elksne, "Lai sauktos par latvieti." Ancītis, "Par latvisko dabas izjūtu," *DVK* 1993, 99–103.

100. Cimdiņa, "Sava kaktiņa, sava stūrīša zemes," 120.

101. Stukuls Eglitis, *Imagining the Nation*, 80–84.

102. Latvian Postal Service web page, http://new.pasts.lv/lv/privatpersonas/filatelija/katalogs/index.html?shop_id=86.

103. According to Tim Unwin and Virginia Hewitt, "The importance of this rural cultural identity to Latvia is also matched, although to a somewhat lesser extent, by the rural imagery depicted on the banknotes of . . . Estonia and Lithuania." Of the banknotes of the new states of Eastern Europe and the former Soviet Union, Unwin and Hewitt report, 80% display human portraits. Paper money is "not only a way of reinforcing internal cohesion and identity, but it is also a way of depicting that identity to the outside world in a

very tangible, and often beautiful, form." Unwin and Hewitt, "Banknotes and national identity in central and eastern Europe," *Political Geography* 20 (2001): 1017, 1018, 1026. See also the Bank of Latvia web page, http://www.bank.lv/eng/main/lvnaud/papnaud/.

104. Students of Bauska High School No. 1, "Letter to Society," *Karogs* 6 (1999): 249–51.

105. Kursīte, *Mītiskais folklorā*, 374.

106. Skaidrīte Albertiņa (ecologist, Latvian SSR Supreme Soviet deputy), interview by the author, Riga, November 3, 1990.

107. Quoted in Mark Bassin, "Turner, Solov'ev, and the 'Frontier Hypothesis': The Nationalist Signification of Open Spaces," *Journal of Modern History* 65 (September 1993): 498, original emphasis.

108. Rasputin, *Siberia on Fire*, 189, 191.

109. Edmunds V. Bunkše, "The Case of the Missing Sublime in Latvian Landscape Aesthetics and Ethics," *Ethics, Place, and Environment* 4, no. 3 (2001): 235–46. Since the National Awakening, Kursīte observes, Latvia's homely landscapes—"the peaceful plains, barely noticeable inclines and moderate hills, calm rivers"—have been identified with defining national characteristics: "non-aggressivity, peaceability, proportionality" (*Mītiskais folklorā*, 374). Latvian landscape aesthetics typically lack religious significance, in marked contrast, for example, to Polish romantics' long tradition of celebrating Bialowieza Forest not only as "primeval" but as the "Great Church of the future religion of the Spirit" (Franklin, "Bialowieza Forest," 1469).

Chapter 3 | Globalizing the Ethnoscape

1. Tālis Tīsenkopfs, "Mūsdienu Latvijas daudzveidīgās identitātes," *Diena*, November 25, 1997.

2. Aivars Ozoliņš, "Savi darbi katram jādara pašam," *Diena*, October 30, 1997.

3. Ilgvars Kļava, "Turpinot diskusiju par identitātēm jeb Kas tie tādi latvieši?" *Diena*, January 22, 1998.

4. Gints Apals, "Mūsdienu Latvijas eiropeiskā identitāte," *Diena*, January 26, 1998.

5. Aigars Daboliņš, "Musu identitāte kā solidaritātes jaunrade," *Diena*, January 7, 1998.

6. Miervaldis Birze, "Vai Latvijai paredzēts Īrijas variants?" *Diena*, February 17, 1998.

7. Valda Čakare, "Atsakieties, reizi tā kā tā būs jāmirst," *Diena*, December 4, 1997.

8. Anda Līce, "Mūsu visu tīrumi," *Diena*, August 8, 1999.

9. Quoted in ibid.

10. Ella Buceniece, "Pret T. Tīsenkopfu, bet vai par pašu problēmu?" *Diena*, December 17, 1997. The Latvian scholar Aija Priedite similarly discusses "two rival models in the public consciousness, both of which are rooted in the history of ideas in Latvia. One model is

open to change, looking with great hope at the process of European integration."
Priedite contrasts this "Westernizing" model with an "ethnic fundamentalism" emphasizing farming as "a lifestyle that is particularly typical and traditional for the Latvian
mentality." Priedite, "Establishment of a Discourse about National Identity in Latvia in
the Late 19th Century and the Early 20th Century," *Philosophical Discourse in Latvia: Humanities and Social Sciences. Latvia* 4, no. 25 (1999): 15–16.

11. According to Strods, of the 500–900 farms in operation between 1970 and 1985, only
150 showed a significant net increase in output, while the remainder had stagnant or diminishing output. *Latvijas Lauksaimniecības vēsture*, 266.

12. In 1988 the average monthly farm wage was 243 rubles in Latvia and 162 rubles for
the Soviet Union as a whole. Van Arkadie and Karlsson, *Economic Survey*, 277.

13. Private plots accounted for 44% of collective farm income in 1970, 25% of milk production, and 18–20% of cattle sales in the 1980s, and 46%–56% of potato production
throughout the Soviet period. This was due in large part to the rampant theft of materials
and labor hours from the collective farms. Boruks, *Zemnieks, zeme un zemkopība*, 375–76;
Misiunas and Taagepera, *The Baltic States: Years of Dependence*, 231, 369.

14. Valdis Liepins, "Baltic Attitudes to Economic Recovery: A Survey of Public Opinion in
the Baltic Countries," *Journal of Baltic Studies* 24, no. 2 (1993): 193. In this cross-Baltic survey, agriculture was ranked first by 68% of respondents in Latvia, 64% in Estonia, and
82% in Lithuania.

15. Ole Norgaard, Lars Johannsen, Mette Skak, and René Hauge Sorensen, *The Baltic
States after Independence*, 2nd ed. (Cheltenham, UK: Edward Elgar, 1999), 122.

16. Latvian State Institute of Agrarian Economics, *Latvijas lauksaimniecība 1998: politika un
attīstība* (Riga: Latvian State Institute of Agrarian Economics, 1999), 107. In 1995, former
collective farms (reincarnated as agricultural companies) accounted for 39% of farmland
in Estonia, 32% in Lithuania, and only 17% in Latvia. William H. Meyers and Natalija Kazlauskiene, "Land Reform in Estonia, Latvia, and Lithuania: a Comparative Analysis," in
Land Reform in the Former Soviet Union and Eastern Europe, ed. Stephen K. Wegren (London:
Routledge, 1998), 90–93.

17. On the Estonian case, see Merje Feldman, "Justice in Space? The Restitution of Property Rights in Tallinn, Estonia," *Ecumene* 6, no. 2 (1999): 165–82. In contrast to the Baltic
states, other countries in Central and Eastern Europe adopted various combinations of
monetary compensation and partial restitution with limits on the size of parcels, and
with varying commitments to distributional justice by granting property to all agricultural workers, regardless of pre-Communist ownership status. With few exceptions, the
collective farm sector retained its dominance throughout the region in the 1990s, rang-
ing in 1995 from around 50% in the Czech Republic and Romania to over 80% in Bulgaria and Slovenia. In Russia, with its long history of statism and communal land tenure
and minimal experience of private family farming, the parliament long refused to formally reinstate private landownership. Despite efforts of the federal government and in-

ternational aid donors, there has been very little breakup of collective farms and minimal growth of private farming. See Csaba Csaki and Zvi Lerman, "Land reform and farm restructuring in East Central Europe and CIS in the 1990s: Expectations and Achievements after the First Five Years," *European Review of Agricultural Economics* 24 (1997): 428–52; Nancy J. Cochrane, "Farm Restructuring in Central and Eastern Europe," *The Soviet and Post-Soviet Review* 21, nos. 2–3 (1994): 319–35; Nigel Swain, "Getting Land in Central Europe," in *After Socialism: Land Reform and Social Change in Eastern Europe*, ed. Ray Abrahams (Providence, RI: Bergahn Books, 1996); Don Van Atta, ed., *The "Farmer Threat": The Political Economy of Agrarian Reform in Post-Soviet Russia* (Boulder: Westview Press, 1993); Wegren, *Land Reform*; and Gene Wunderlich, ed., *Agricultural Landownership in Transitional Economies* (Lanham, MD: University Press of America, 1995).

18. Meyers and Kazlauskiene, "Land Reform in Estonia, Latvia, and Lithuania," 98. The corresponding figures are 134,600 in Lithuania and 13,500 in Estonia.

19. Paul Sando, "Latvian Agriculture After Communism: Restructuring and Privatization" (Ph.D. diss., Indiana State University, 1996), 24.

20. Boruks, *Zemnieks, zeme un zemkopība*, 393.

21. Anatol Lieven, *The Baltic Revolution: Estonia, Latvia, Lithuania and the Path to Independence* (New Haven: Yale University Press, 1993). The economists Brian Van Arkadie and Mats Karlsson concur: "The family farms introduced in the Baltic states so far are of greater importance as a cultural and national manifestation than as a means for improving productivity." Van Arkadie and Karlsson, *Economic Survey*, 282. Stephen K. Wegren argues similarly that "cultural continuities" are critical in explaining the lack of mass decollectivization in Russia. Wegren, "Rural Reform and Political Culture in Russia," *Europe-Asia Studies* 46, no. 2 (1994): 215–41. The carryover into the post-Soviet period of pre-Soviet Russian cultural norms of collectivism and egalitarianism provides a striking contrast to the carryover of agrarian nationalism in Latvia. Another revealing contrast is provided by Gerald W. Creed, who notes that Bulgarian villagers welcomed the depeasantization wrought by collectivization and, consequently, resisted decollectivization as a form of re-peasantization. Under socialism, the "expansion of non-agricultural occupations and the distinction between professional (cooperative) and subsistence (household) production provided villagers personal distance from the culturally demeaning association with agriculture *as an occupation*. Few villagers in the late 1980s, especially those of working age, considered themselves farmers; their identities were linked instead to non-agricultural jobs in small enterprises or village administrative posts, bringing them closer to the proletarian ideal of communism. Even those employed by cooperative farms saw themselves as 'tractor drivers' or 'combine operators rather than as farmers or peasants." Gerald W. Creed, "The Politics of Agriculture: Identity and Socialist Sentiment in Bulgaria," *Slavic Review* 54, no. 4 (1995): 843–68. My anecdotal evidence suggests that in Latvia many former kolkhoz employees, from drivers to skilled technical professionals, did not share this reluctance to return to the personal status (and heavy workload) of individual farming.

22. Talis Tīsenkopfs, "Rurality as a Created Field: Towards an Integrated Rural Development in Latvia?" *Sociologia Ruralis* 39, no. 3 (1999): 416.

23. Valentina Mičūrova (deputy director, Ministry of Agriculture Reclamation and Land Use Unit), interview by the author, Riga, October 23, 1998. If Mičūrova and Tīsenkopfs are correct in their assessments of new Latvian private farmers' attitudes, then the Latvian experience differs significantly not only from Bulgaria's but even from Estonia's. According to the authors of a 2001 study, radical decollectivization in Estonia was the product of a utopian vision of returning to the idyll of the interwar family farm, imposed by out-of-touch urban intellectuals on an unwilling rural population. Instead of 1920s-style family farming, most collective farm employees would have preferred to maintain some mix of large- and small-scale production, involving a democratization of the collectives along with expanded private plot farming. In Estonia, this study concludes, the perverse marriage of neoliberal economic prescriptions with "family-farm orthodoxy" has failed to produce a viable family-farm sector. Instead, most small farms have collapsed or teeter on the verge of subsistence, and much of the rural population is trapped in a "vicious circle of pauperization." Ilkka Alanen, Jouko Nikula, Helvi Poder, and Rein Ruutsoo, eds., *Decollectivization, Destruction and Disillusionment: A Community Study in Southern Estonia* (Aldershot, UK: Ashgate, 2001), 257.

24. Farm incomes plummeted from from 722 lats per capita in 1995 to 482 lats in 1998. Latvian State Institute of Agrarian Economics, *Latvijas lauksaimniecība 1998*, 40.

25. Ibid., 49.

26. Ibid., 23, 40.

27. Central Statistical Bureau, *Latvia: Chief Statistical Indicators 1999*, 11.

28. In 1997 the average monthly wage in the agrarian sector (excluding fisheries) was 63 lats, as compared to an overall average of 88 and industrial average of 97. Central Statistical Bureau, *Statistical Yearbook of Latvia 1998*, 99.

29. In the second half of the 1990s, agricultural support levels in Latvia, calculated in terms of production subsidy equivalents, lagged considerably behind those in Lithuania and even Estonia, vaunted for its commitment to rapid and radical liberalization. Latvian State Institute of Agrarian Economics, *Latvijas lauksaimniecība 1998*, 112.

30. Tisenkopfs, "Rurality as a Created Field," 427.

31. Central Statistical Bureau, *Latvia: Chief Statistical Indicators 1999*, iii, 54.

32. Latvian State Institute of Agrarian Economics, *Latvijas lauksaimniecība 1998*, 116.

33. Arvīds Bondars, "Latvijas lauksaimniecība ir gandrīz pilnībā sagrauta," *Dienas bizness*, March 15, 1996.

34. See, for example, Ansis Dannenbergs, "Un atkal jaunā kolhozā?" *Diena*, September 30, 1992; Oskars Grīgs, "Laukos—nācijas pamats," *Diena*, November 20, 1992; Vilnis

Zariņš, "Latvijai vajadzīgi stipri zemnieki," Diena, July 29, 1993; and Akvilina Liepiņa, "Latvijas zeme pošas uz tirgu," Diena, November 20, 1993.

35. Ainārs Dimants, "Jābruģē ceļš mūsu ražotājiem," Diena, March 2, 1993.

36. Dannenbergs, "Un atkal jaunā kolhozā?"

37. Dimants, "Jābruģē ceļš."

38. Tisenkopfs, "Rurality as a Created Field," 416.

39. Liepiņa, "Brīva zemes tirgus vēl nebūs," Diena, August 26, 1994.

40. Anita Smoļenska, "Brīvs zemes tirgus un Latvijas izpārdošana," Diena, June 21, 1994.

41. Much as in Latvia, McCarthy argues that for the Wise Use movement of the 1980s and 1990s, the "myth of the Jeffersonian smallholder was by far this alliance's most potent rhetorical trope." McCarthy, "First World Political Ecology," 1291, 1295.

42. Katherine Verdery and Martha Lampland note the same phenomenon for Romania and Hungary, respectively. According to Verdery, in land disputes during decollectivization in Romania, the two salient themes were kinship and work: "People deserve land because it belonged to their ancestors, because they or their parents worked it, or because they are better able than others to work the land now and in the future." This emphasis on work derives not "chiefly from socialism's emphases on work and production; rather it is rooted in pre-collective village notions about personhood, or the self, as something constructed through labor and possession." Verdery, "Nationalism, Postsocialism, and Space," 87. Similarly, Lampland argues that the "centrality of labor to one's identity and honor within the community" in interwar Hungarian villages was reinforced under socialism by "policies attempting to separate out production and property, [whereby] the identity of workers defined by their actions as laborers was strengthened." Lampland, The Object of Labor, 336–37.

43. Liepiņa, "Latvijas zeme pošas uz tirgu."

44. Stukuls Eglitis, Imagining the Nation, 87.

45. Ilze Arklina, "Artistic Controversy Erupts around President's Memory," Baltic Times, November 1–7, 2001. Ultimately a different design was implemented, minus the field of rye.

46. U. Šmits, "Kārlis Ulmanis—Latvijas ieverojamākais valstsvīrs," Lauku Avīze, September 5, 1997. The conference was co-sponsored by the Latvian Academy of Sciences, the University of Latvia Institute of History, the Latvia Farmers' Union, the Kārlis Ulmanis Memorial Foundation and the Riga Latvian Association.

47. Boruks, Zemnieks, zeme un zemkopība, 252, 290. Boruks compares the Ulmanis regime to the East Asian developmental state. A similar assessment is offered by the agronomist E. Bunga in "Kārlis Ulmanis—agronoms, tautsaimnieks, politiķis," Lauku Avīze, September 2, 1997.

48. Strods, *Latvijas Lauksaimniecības vēsture*, 174.

49. Egils Līcītis, "Kārlis Ulmanis šodien būtu aktīvā vietā un smagā darbā," *Lauku Avīze*, May 20, 1997.

50. Anon., "Jāstrādā un jātaupa," *Druva*, July 31, 1998.

51. Aivars Tarvīds, "Graudi un pelavas," *Vakara Ziņas*, August 26, 1998.

52. Dzintars Korns, Līga Krustkalne and Ansis Zunde, "Uguns & grēka suģestija," *Filosofija* 1 (1998): 231–36.

53. Aivars Tabūns, "Musu mainīgās un daudzveidīgas identitātes," *Diena*, January 17, 1998.

54. Apals, "Mūsdienu Latvijas eiropeiskā identitāte."

55. John Lloyd, "Squeezed Between Two Blocs," *Financial Times*, November 19, 1994.

56. Teodors Tverijons (president of Latvia's Commercial Banks Association), quoted in Gary Peach, "Building Financial Bridge Between East, West," *Baltic Times*, December 12–18, 2002.

57. Guntis Valujevs, "Racionāla ekonomika un silueti miglā," *Diena*, April 1, 1993.

58. Ibid.

59. In 1998, Latvia's population density was 38.1 persons per square kilometer for the country as a whole, and ranged from 20 to as low as 5.7 in the provinces. Central Statistical Bureau, *Statistical Yearbook of Latvia 1998*, 56. By comparison, average population density in the EU-15 in 1998 was 117 per square kilometer, with highs of 334 and 382 for Belgium and the Netherlands, respectively. Excluding the partially inhospitable Nordic countries, the most sparsely populated member state was Ireland with 54. Eurostat, *Eurostat Yearbook 2000* (Luxembourg: Office for Official Publications of the European Communities, 2000), 17.

60. Miglavs, "Lauki nav tikai lauksaimnieciskā ražošana," *Latvijas Vēstnesis*, November 21, 1997; Miglavs, 1999 interview.

61. Aigars Štokenbergs, "Laukus glābs Rīga katrā mazpilsētā," *Diena*, November 5, 1997.

62. Dannenbergs, "Un atkal jaunā kolhozā?"

63. Grīgs, "Laukos—nācijas pamats."

64. Jānis Freimanis, "Mazo valstu lielā rūpe," *Diena*, June 18, 1992.

65. Boruks, *Zemnieks, zeme un zemkopība*, 414.

66. Miglavs, 1999 interview.

67. Latvijas lauku atbalsta asociācijas aicinājums, "Gatavosimies Latvijas lauksaimnieku kongresam!" (unpublished document, 1998).

68. Bunkše, "God, Thine Earth Is Burning," 204.

69. On agricultural decline, land abandonment, and national identity in post-Communist Estonia, Georgia, and Poland, see Tim Unwin, "Contested Reconstruction of National Identities in Eastern Europe: Landscape Implications," *Norsk geografisk Tidsskrift* 53 (1999): 113–20.

70. Līce, "Mūsu visu tīrumi."

71. Until a sharp downturn in 2002 following the opening of a rival Russian port at Primorsk, Ventspils serviced the lion's share of total cargo turnover in Latvia (73% by volume in 1997), primarily oil and oil products from Russia. Oil and oil products comprised 61% by volume of cargoes loaded at Latvian ports in 1997. Central Statistical Bureau, *Statistical Yearbook of Latvia 1998*, 257–85. In the 1990s, Ventspils ranked first "among all ports in EU candidate countries in terms of cargo turnover, according to Eurostat research on maritime development in EU candidate countries for the five years leading to 2000." *Baltic Times*, October 10–16, 2002.

72. Baiba Rulle, "Lembergs: Valsts vadītājiem trūkst pasaules redzējuma un izglītības," *Diena*, May 31, 1999. Lembergs and other representatives of the transit trade have self-consciously sought to appropriate Valdemārs's legacy. A collection of Valdemārs's writings on maritime topics (as well as the complete text of the 1867 imperial Law on Naval Schools and summaries in English and Russian of Valdemārs's "National Seafaring Development Program") was published in 1995 by the Latvian Seafaring Fund (Lūsis and Šīmanis, *Krišjānim Valdemāram* 170). The project was funded by maritime firms, trade associations, and state agencies, including the Association of Latvian Shipowners, Latvian Shipping, the port authorities of Ventspils, Liepaja, and Riga, and the Latvian Maritime Administration. Third in the list of sponsors is the transshipment giant Ventspils Nafta, followed by two smaller Ventspils firms and the Ventspils Commercial Port.

Chapter 4 | Returning to a Postproductivist Europe

1. Miglavs, 1999 interview.

2. The program began with the acronym PHARE, or Poland/Hungary Assistance for Restructuring the Economy.

3. See Sven Arnswald, *EU Enlargement and the Baltic States: the Incremental Making of New Members* (Helsinki and Bonn: Ulkopoliittinen instituutti and Institut für Europäische Politik, 2000).

4. In 1998, average agricultural productivity in terms of value added per worker in the ten CEEC candidate countries (including Bulgaria and Romania) was only 11% of the EU level. European Commission Directorate for Agriculture, *Summary Report: Agricultural Situation and Prospects in the Central and Eastern European Countries* (Luxembourg: Office for Official Publications of the European Communities, 1998), 5. In 1995, Latvian GDP at Purchasing Power Standard was only 23% of the EU average. Dzintra Bungs, *The Baltic States:*

Problems and Prospects of Membership in the European Union (Baden-Baden: Nomos Verlagsgesellschaft, 1998), 21.

5. Launched in the 1960s, the CAP reflected the postwar concerns of boosting agricultural productivity and ensuring the security of food supplies and stability of markets. Its primary policy tool has been price supports targeted at particular commodities. This mechanism has been so successful in achieving the CAP's original goals that production surpluses began to appear in the 1980s. It has been less successful, however, in promoting rural development more broadly, and so a number of programs have been implemented—the Structural Funds, the Cohesion Fund and most recently the Less Favored Areas program—to promote rural development in disadvantaged regions.

6. Clive Potter and Philip Goodwin, "Agricultural Liberalization in the European Union: An Analysis of the Implications for Nature Conservation," *Journal of Rural Studies* 14, no. 3 (1998): 289.

7. Jim Dixon, "Implementation of Agri-environment Regulation 2078," in McCracken et al., *Farming on the Edge*, 158.

8. See, for example, Beaufoy et al., *The Nature of Farming*.

9. Nowadays agriculture accounts for less than 2% of GDP and just over 5% of employment in the EU. European Commission Directorate General VI/A1, *Towards a Common Agricultural and Rural Policy for Europe* (Luxembourg: Office for Official Publications of the European Communities, 1997), 5.

10. These measures included direct subsidies and a voluntary set-aside program whereby land must be fallowed, wooded, or used for nonagricultural purposes such as recreation and conservation. The MacSharry reforms of 1992 reduced price supports, imposed mandatory set-asides on larger producers, and strengthened the environmental program by expanding the range of subsidized activities and making the program permanent and binding on all member states. A Species and Habitats Directive was issued in 1992, establishing the Natura 2000 network of "special protection areas" for the study and conservation of biodiversity. See Buller et al., *Agri-environmental Policy*; and Martin Whitby, ed., *The European Environment and CAP Reform: Policies and Prospects for Conservation* (Wallingford, UK: CAB International, 1996).

11. European Commission, *Towards a Common Agricultural and Rural Policy*, 39. In part, this new view was responding to an already changed reality: by the late 1980s, "one-third of farm houselands obtained more than 50% of their income from off-farm sources." Brian Ilbery and Ian Bowler, "From Agricultural Productivism to Post-Productivism," in *The Geography of Rural Change*, ed. Ilbery (Edinburgh Gate, UK: Longman, 1998), 75.

12. John Bryden, *Towards an Agenda for Sustainable Rural Development in Europe: A Discussion Paper* (Aberdeen, UK: Arkleton Centre for Rural Development Research, 1998); Erwin Stucki, *Balanced Development of the Countryside in Western Europe: The Conditions for a Balanced Long-term Development of the Countryside Respecting the Natural Equilibrium*, Nature and environment, no. 58 (Strasbourg: Council of Europe Press, 1992); World Wide Fund for Na-

ture, *A New European Community Policy—Sustainable Regional Development* (Germany: WWF-Deutschland, 1997).

13. Quoted in Buller et al., *Agri-environmental Policy*, 3.

14. Potter and Goodwin, "Agricultural Liberalization," 294.

15. Baker et al., *The Politics of Sustainable Development*, 4.

16. Ibid., 5.

17. Takacs, *The Idea of Biodiversity*, 75.

18. Mikael Skou Anderson, "Ecological Modernization or Subversion? The Effect of Europeanization on Eastern Europe," *American Behavioral Scientist* 45, no. 9 (2002): 1396. According to Anderson, this aid has not amounted to anything resembling a "green Marshall plan for Europe." Instead, "the East European countries have had to finance most environmental investments on their own. In the CEE countries, the foreign assistance has generally not exceeded 5% of annual investments, not even in Poland and Hungary, which have been the largest recipients, and the contribution is most cases closer to 1% to 2%. Only in the Baltic states has the assistance reached a share of 30% to 35% of the investments, mainly thanks to aid from Nordic countries" (ibid., 1402). See also Robert O. Keohane and Marc Levy, eds., *Institutions for Environmental Aid* (Cambridge: MIT Press, 1996). For an overview, see JoAnn Carmin and Stacy D. VanDeveer, eds., *EU Enlargement and the Environment: Institutional Change and Environmental Policy in Central and Eastern Europe* (London: Routledge, 2005). This volume is a reprinting of a special issue of *Environmental Politics* 13, no. 1 (Spring 2004).

19. "Almost all streams and rivers of the Baltic republics eventually carry their water and attendant heavy load of pollutants into the Baltic Sea," which thirty years ago "was vying for the title of the most polluted sea in the world." Juris Dreifelds, "The Environmental Impact of Estonia, Latvia and Lithuania on the Baltic Sea Region," in *Environmental Security and Quality after Communism: Eastern Europe and the Soviet Successor States*, ed. Joan DeBardeleben and John Hannigan (Boulder: Westview Press, 1995), 155. See also Robert G. Darst Jr., "Bribery and Blackmail in East-West Environmental Politics," *Post-Soviet Affairs* 13, no. 1 (1997): 42–77.

20. Ministry of Environmental Protection and Regional Development (MEPRD), *National Environmental Policy Plan for Latvia* (Riga: MEPRD and the Ministry of Housing, Physical Planning, and Environment of the Netherlands, 1995), 1.

21. Ministry of Environmental Protection and Regional Development, *National Report on Biological Diversity: Latvia* (Riga: MEPRD and United Nations Development Programme, 1998), 5.

22. From 1995 to 1997 the ministry was headed by the radically productivist Alberts Kauls. Formerly a legendarily successful kolkhoz director and Gorbachev adviser, even in the 1990s Kauls remained less interested in EU accession than in restoring Soviet-era trade links and methods of governance. He achieved dubious renown by paying im-

promptu visits to Moscow to arrange barter deals with old cronies and dispatching government officials to the countryside to enforce harvesting.

23. Andris Miglavs (director, Latvian State Institute of Agrarian Economics), interview by the author, Riga, July 19, 1999.

24. Latvian State Institute of Agrarian Economics, *Latvijas lauksaimniecība 1998*, 118.

25. The EU launched two new programs for channeling rural development assistance to the eastern aspirants: ISPA (Instrument for Structural Policies for Pre-Accession) and SAPARD (Special Action for Pre-Accession measures for Agriculture and Rural Development).

26. Ministry of Agriculture, "Latvijas SAPARD programmas projekts" (unpublished document, May 1999), 21.

27. Tisenkopfs observes: "The Ministry of Agriculture is focused on strengthening the agricultural production sector, but it also handles a subsidy program for diversification into non-traditional farming and off-farm activities. The Ministry of Finance encourages small-scale enterprise development in rural areas through various tax concessions. The Ministry of Economy operates a Regional Fund that provides grants and interest subsidies for private and public sector investments in less-developed areas and is implementing the SME [small and medium-sized enterprise] Development Program. The Ministry of Environmental Protection and Regional Development is involved in rural development through supervision and the allocation of funding for capital expenditures to local governments, thus assisting them in the preparation of territorial plans and development strategies. The State Employment Service runs an active employment policy through its branch offices in rural areas." Tisenkopfs, "Rurality as a Created Field," 425–26.

28. Heinrich Hick (member, EU delegation to Latvia), interview by the author, Riga, July 27, 1999. See also Torben Nielsen (team leader, PHARE project for institutional development in support of Latvian agriculture), interview by the author, Riga, July 7, 1999; and Aivars Lapiņš (director, Ministry of Agriculture Department of Agricultural Strategy and Cooperation Rural Development Unit), interview by the author, Riga, July 8, 1999.

29. Latvian State Institute of Agrarian Economics, *Latvijas lauksaimniecība 1998*, 118.

30. Steven Sampson discusses the art of "project-speak" in "The Social Life of Projects: Importing Civil Society to Albania," in *Civil Society: Challenging Western Models*, ed. Chris Hann and Elizabeth Dunn (London: Routledge, 1996).

31. Valts Vilnīts, interview by Egils Zirnis, *Diena*, November 20, 1991.

32. "Mires (together with peat fields) occupy 5.6% of the territory of Latvia. 70% of them are relatively undisturbed by human activities." MEPRD, *National Report on Biological Diversity*, 7.

33. IUCN East Europe Program, *The Wetlands of Central and Eastern Europe* (Gland, Switzer-

land: IUCN, 1993), 22; MEPRD, *National Report on Biological Diversity*, 12. In 1995, according to Latvian statistics, there were between 900 and 1,300 nesting pairs of black storks in Latvia, as compared to 300 to 400 in Lithuania, one in Denmark, and none in Sweden or the Netherlands. Environmental Protection Club of Latvia-USA chapter, *Open Letter* 13–14 (1991–1992).

34. World Wide Fund for Nature, *Latvia's Natural Heritage at the Crossroads* (Riga: WWF Latvia, ND), 10.

35. IUCN East Europe Program, *The Wetlands of Central and Eastern Europe* (Gland, Switzerland: IUCN, 1993), 37.

36. Gunita Ozoliņa, "Pasaulslaveni pētnieki sajūsminās par Latvijas purviem," *Diena*, July 15, 1998.

37. Interview by Zirnis.

38. Jānis Priednieks (chairman, Latvian Fund for Nature), interview by the author, Riga, July 9, 1999.

39. Maruta Kaminska (ecosystem inspector, Liepāja regional environmental agency), interview by the author, Liepāja, January 13, 1998.

40. Imants Baumanis (Silava Institute of Forest Science), meeting of EU-approximation task force on agriculture and forestry, Riga, December 23, 1998.

41. MEPRD, *National Report on Biological Diversity*, 5.

42. MEPRD, *National Environmental Policy Plan*, 3.

43. Interview by Zirnis.

44. "Prece—pirmatnīgs mežs un pļava," *Diena*, May 3, 1997.

45. Ministry of Agriculture, "Latvijas SAPARD programmas projekts," 24.

46. Ministry of Agriculture, "Latvijas lauku attīstības plāns Eiropas Savienības pirmsiestāšanās pasākumiem lauksaimniecības un lauku attīstībai (saīsināts), 2000.–2006. g." (unpublished document, August 1999), 22.

47. "Latvia: Heartland of the Baltics," advertisement, *Financial Times*, July 6, 1998. Promoters in Estonia have similarly drawn on the biodiversity narrative, as in this excerpt from *Estonia for Tourists*: "The beauty of animate and inanimate nature with its wide variety, and the preservation of naturalness—this is the greatest treasure of Estonia. Land and sea, jagged coastline and more than a thousand islands, high bank and picturesque dunes, stone-covered areas side by side with patches of fertile soil, hills and primeval valleys, springs and a thousand lakes, karst areas and marshland, extensive territories covered with forest, flora extremely rich in species, habitats of waterfowl, landscapes long forgotten in many European countries." Quoted in Tim Unwin, "Rurality and the Construction of Nation in Estonia," in *Theorising Transition: The Political Economy of Post-Communist Transitions*, ed. John Pickles and Adrian Smith (London: Routledge, 1998), 299.

48. Andris Junkurs (project manager, Northern Kurzeme Ecotourism Development Plan), interview by the author, Liepāja, February 18, 1999.

49. Junkurs, presentation at the "Amber Trail" ecotourism seminar, Vērgale, February 16, 1999.

50. Gunta Strēle (Bārtava director), interview by the author, Kalēti, February 11, 1999.

51. Euroregions are established through the EU's INTERREG and CBC PHARE programs, launched in 1990 and 1994, respectively, to promote economic development and cross-border cooperation.

52. World Wide Fund for Nature, A History of WWF (Gland, Switzerland: WWF, 1998), 1.

53. World Wide Fund for Nature, WWF Project 4568: Conservation Plan for Latvia, Final Report (Riga: LU Ekoloģiskais centrs, 1992), 7.

54. WWF-Latvia was founded as a Representative Office of WWF-International, under the legal auspices of WWF-Switzerland. Because Switzerland is not an EU member state, WWF-Latvia was also barred from receiving EU funds. According to Latvian law, WWF-Latvia was not recognized as a Latvian NGO, and was thereby ineligible for tax-exempt status for domestic fund-raising, as well as for competing for Latvian government funds, such as grants from the Environmental Protection Fund. In 2002, WWF-Latvia was reorganized as a nonprofit public foundation under Latvian law, thus making it eligible to compete for both domestic and EU funding.

55. Steward T. A. Pickett, V. Thomas Parker, and Peggy L. Fiedler, "The New Paradigm in Ecology: Implications for Conservation Biology Above the Species Level," in Conservation Biology: The Theory and Practice of Nature Conservation, Preservation and Management, ed. Fiedler and Subodh K. Jain (New York: Chapman and Hall, 1992), 74. On this paradigm shift, see Stephen Budiansky, Nature's Keepers; The New Science of Nature Management (New York: Free Press, 1995).

56. Mārtiņš Rēķis (WWF-Latvia rural development project director), interview by the author, Riga, August 4, 1999.

57. For a critique of this strategy of "green developmentalism," pursued by WWF as well as other mainstream conservationist organizations, academic scientists and international organizations such as IUCN, the World Bank and the OECD, see Kathleen McAfee, "Selling Nature to Save it? Biodiversity and Green Developmentalism," Environment and Planning D: Society and Space 17 (1999): 133–54.

58. Rēķis, 1999 interview.

59. United Nations Conference on Environment and Development, Agenda 21 and the UNCED Proceedings (New York: Oceana, 1992), 1.

60. MEPRD, National Report on Biological Diversity, 29.

61. Anon., "Par teritoriālplānošanas reformu," Diena, April 23, 1996.

62. Maira Ādmine, "*Lauku ceļotājam piecu gadu jubileja,*" *Dienas bizness,* December 9, 1998.

63. Quoted in Nick Coleman, "Wild Beauty Amidst the Beast of Rural Poverty," *The Baltic Times,* June 13–19, 2002. Coleman reports that tourism contributed only 1.5% to Latvia's GDP in 2001.

64. Ilbery and Bowler, "From Agricultural Productivism," 100.

Chapter 5 | "Masters in Our Native Place"

1. Kenneth R. Olwig, "Reinventing Common Nature: Yosemite and Mount Rushmore— A Meandering Tale of a Double Nature," in Cronon, *Uncommon Ground,* 380.

2. Nash, *Wilderness and the American Mind;* Alfred Runte, *National Parks: The American Experience* (Lincoln: University of Nebraska Press, 1979).

3. Mark David Spence, *Dispossessing the Wilderness: Indian Removal and the Making of National Parks* (New York: Oxford University Press, 1999), 4.

4. In addition to Spence, see also Robert H. Keller and Michael E. Turek, *American Indians and National Parks* (Tucson: University of Arizona Press, 1998); and Philip Burnham, *Indian Country, God's Country: Native Americans and the National Parks* (Washington, DC: Island Press, 2000).

5. Jane Carruthers, "Creating a National Park, 1910 to 1926," *Journal of Southern African Studies* 15, no. 2 (1989): 188–216.

6. Roderick Neumann, "Ways of Seeing Africa: Colonial Recasting of African Society and Landscape in Serengeti National Park," *Ecumene* 2, no. 2 (1995): 149–69.

7. Terence Ranger, "Whose Heritage? The Case of the Matobo National Park," *Journal of Southern African Studies* 15, no. 2 (1989): 218–19. See also Ranger, *Voices from the Hills: Nature, Culture and History in the Matopos Hills of Zimbabwe* (Bloomington: Indiana University Press, 1999); Adams and McShane, *Myth of Wild Africa.*

8. Carruthers, "Creating a National Park," 207.

9. Interview by the author, Riga, March 11, 1999.

10. Melluma, address to the Gauja National Park zoning seminar, Sigulda, December 9, 1998. This phenomenon presents an interesting topic for comparative research, though one beyond the scope of this study.

11. Melluma, "Par nacionālo parku 'Gauja,'" *DVK* 1974, 123–26.

12. Guntis Eniņš, "Kādam būt Gaujas nacionālajam parkam," *Padomju jaunatne,* March 7, 1973. The founders of Estonia's Lahemaa National Park had a similar approach, as Rob Smurr reports. According to the park's "founding father," Jaan Eilart, "'The park had everything to do with Estonians and their culture and nothing to do with nature.' But the

Estonian culture Eilart sought to protect was a traditional peasant culture intimately tied to the land; thus, despite his contention to the contrary, the park was about nature too. It was a place where the natural and built environments served as congenial complements to one another." Smurr, "Nationalizing Nature," 25.

13. Melluma, "Par nacionālo parku."

14. Moritzholm reserve was established eight years before Germany's first nature reserve, established in the Luneburg Heath in Lower Saxony in 1920. Williams, "The Chords of the German Soul," 351.

15. Pryde, "The Environmental Basis for Ethnic Unrest," 11.

16. N. Austrins, "Mežs un koki 'Latvju Dainās'," *Meža Dzīve* 1926, 8–9: 247. See also, for example, Edgars Teidoffs, "Daiļuma kopšana Latvijas mežos un laukos," *Meža Dzīve* 1938, 5: 235–37.

17. Edvarts Jansons, "Dabas pieminekļi Latvijā," in *Latvijas zeme, daba un tauta*, vol. 2, ed. N. Malta and Galenieks (Riga: Valters un Rapa, 1936), 322. For another writer, man-made objects even seemed to come mind first, before natural ones: "By 'monuments of nature' we understand . . . not only ancestral buildings, homesteads and other monuments, but also all of nature's wonders: blessed oaks, old trees and rare plants and animals." Retelis, "Dabas aizsardzība Latvijā," *Meža Dzīve* 1932, 83–84: 3089–91.

18. Weiner, *Models of Nature*. On how the biological and agricultural sciences fell victim to Stalinist proponents of *partiinost'* and radically utilitarian "Prometheanism," see Graham, *Science in Russia*; Joravsky, *The Lysenko Affair*; and Josephson, *Totalitarian Science*.

19. Aija Melluma, *Latvijas PSR aizsargājamas dabas teritorijas* (Riga: Zinātne, 1979), 18.

20. Weiner, *A Little Corner of Freedom*, 393.

21. Melluma, *Latvijas PSR aizsargājamās dabas teritorijas*, 15.

22. Ibid., 11.

23. Melluma, "Metamorphoses of Latvian Landscapes," 56.

24. Melluma, "Kā veidosim?" *Literatūra un Māksla*, April 11, 1980.

25. Indra Čekstere, "GNP etnogrāfiskās vērtības, to apzināšana un aizsardzība," *Mežsaimniecība un Mežrūpniecība*, 1983, no. 6: 47.

26. When I asked Čekstere whether her informants were willing to talk about their lives, she said they spoke "entirely openly, without fear," speculating that they trusted her because of her youth. She added: "I had permission to record people's recollections. I couldn't record them in my reports, but nobody inspected my notes." Indra Čekstere, GNP information officer; interview by the author, Sigulda, January 20, 1998. Excerpts from Čekstere's field notes are published in Čekstere, "Vasara Vaidavā," *Meža Dzīve*, 1993, no. 7: 17–23.

27. Čekstere, 1998 interview.

28. Čekstere, "GNP etnogrāfiskās vērtības," 48–50.

29. G. Skriba, "Gaujas nacionālā parka 20 darba gadi," *Meža Dzīve*, 1993, no. 7: 3–4.

30. See State Forest Service and Holsteinborg Consult, *Management Plan for Gauja National Park: Inception Report* (unpublished document, 1998).

31. Valdis Pilāts (GNP senior ecologist), interview by the author, Sigulda, January 28, 1999. Restricted access to scientific information and maps was a common feature of Communist regimes. "In most cases, data had to be copied by hand according to the needs of each of the specific research projects. Ordinary people and researchers were not allowed access to copying and reproduction machines. Since data collection was always specific to a particular research project, compilation of systematic and complete databases was difficult. Moreover, since no 'information market' existed, there was only restricted development of the tools of information storage, management and flow." Pavlinek and Pickles, *Environmental Transitions*, 233.

32. IUCN, *Guidelines for Protected Area Management Categories* (Gland, Switzerland: IUCN, 1994), 12, 19.

33. Jānis Strautnieks (GNP director), interview by the author, Sigulda, August 13, 1999.

34. Ibid.

35. Franklin "Bialowieza Forest," 1480; Strautnieks, "Ar nodilušām pastalām," *Meža Dzīve*, 1993, no. 7: 2. Franklin argues that Poland's Bialowieza National Park served to "atone for pollution elsewhere in Poland at a time when Poland [was] attempting to meet the Copenhagen Criteria for European accession."

36. Valdis Pilāts, "Ko tūrists grib redzēt Gaujas nacionālajā parkā," *Meža Dzīve*, 1993, no. 7: 27–28.

37. Aase Ostergaard (GNP management plan project leader), interview by the author, Sigulda, January 20, 1999.

38. Melluma, August 16 interview.

39. Melluma, March 11 interview.

40. The first nature reserve on Latvian territory, and one of the earliest reserves in the Russian empire, was established in 1912 on Moritzholm Island (in Latvian, Moricsala).

41. Dundagas pagasta padome and SIA "Blezūrs—Konsultāciju birojs," *Ziemeļkurzemes izpētes projekts: noslēguma ziņojums* (unpublished document, 1999), 15.

42. Arvīds Plaudis, "Šie ciemi dzīvos īpatnēju dzīvi," *Elpa*, October 7, 1994.

43. Gunārs Laicāns (Dundaga township mayor), telephone interview by the author, August 12, 1999.

44. Gundega Blumberga, "Viens liels dabas parks un—nekādu rezervātu!" *Līvli* July/August 1996.

45. Elmārs Pēterhofs (Slītere Reserve director), interview by the author, Dundaga, April 4, 1999.

46. Egils Jucēvičs, "Slītere šodien un rīt," *Kurzemes Ekspresis*, March 21, 1997.

47. I. Andiņš, "Parku—putniem, rezervātu—cilvēkiem?" *Lauku Avīze*, May 6, 1997.

48. Jucēvičs, "Slītere šodien un rīt." At the time Latvia had two national parks, the second having recently been established at Ķemeri, an ecologically significant wetland area along the Gulf coast just west of Riga. Poorly developed and severely underfunded, however, Ķemeri was at the time only a "paper park," existing more in name than in fact.

49. O. Zakrisons, A. Granstrēms and Dž. Šimmels, "Slīteres Valsts rezervāts Zviedrijas Lauksaimniecības universitātes Meža ekoloģijas katedras speciālistu vērtējumā," *Meža Dzīve*, 1993, no. 5: 17–19.

50. Ministry of Environmental Protection and Regional Development, "Management plan for Sūr Mer National Park" (unpublished document, n.d.), 9. Sūr Mer (Great Sea), the Livonian name for the Baltic Sea, was considered as an alternative name for the proposed park.

51. Andiņš, "Parku—putniem."

52. Gunita Ozoliņa, "Kāpēc kurzemnieki noraida nacionālā parka modeli," *Diena*, November 23, 1997.

53. Skaidrīte Lapiņa (Ance township mayor), interview by the author, Ance, July 21, 1999.

54. Laimdota Sēle, "Kurzemes jūrmala—dzīvibas zeme vai rezervāts?" *Ventas balss*, February 20, 1997.

55. G. Stalte, "Gadskārtējā līvu konference dziesmotā garā," *Banga*, May 9, 1997.

56. Ibid. Consider the following similar take on official celebrations of cultural heritage: "A Latvian English-language teacher who visited the United States in 1993 as part of a US government exchange program recounted to me that her program's visit to a Native American reservation reminded her of the darkest days of Soviet rule, when Latvian folk culture was paraded for visiting foreign dignitaries as evidence of the government's love of indigenous cultures. She reported that when she attended a staged Native American ritual performance during her reservation visit she began sobbing because she so identified with the performers." John Ginkel, "Identity Construction in Latvia's 'Singing Revolution': Why Inter-ethnic Conflict Failed to Occur," *Nationalities Papers* 30, no. 3 (2002): 421.

57. Maija Rēriha (Kolka township mayor), interview by the author, Kolka, May 5, 1999.

58. Jānis Trops, "Kurzemes krasta krustceles," *Diena*, March 18, 1997.

59. Sēle, "Kurzemes jūrmala."

60. Edgars Sīlis (Livonian Coast executive director), interview by the author, Riga, June 4, 1999.

61. Rēriha, 1999 interview. In the version in my possession, the section on "heritage environments" does call for negotiating with landowners and finding "ways of subsidising the traditional land use forms," as cited above. However, the drafters appear to have made a tactical error by preceding these statements with an appeal to biodiversity, thereby suggesting the primacy of biological over cultural values: "The traditional single farmsteads and small villages (both inland and on the coast) in a territory of a continuous forest massive is a significant element for the maintenance of biological variety in the nature: birds, bats, insects, plants and other forms of life enrich the landscape of Sūr Mer National Park." MEPRD, unpublished document, n.d., 9.

62. Rēriha, 1999 interview by the author.

63. Sēle, "Vai jūrmalas balsis."

64. Andiņš, "Parku—putniem."

65. Laicāns, 1999 interview. Stuart Franklin describes a similar dynamic in the Bialowieza case, where local residents resisted efforts by Polish scientists in the 1990s to create a national park, seeing the current emparkment as reminiscent of the Nazi plan to ethnically cleanse the Polish-Belorussian border zone. "Between June and August 1941 thousands of Belorussian farmers were driven out of their forest villages as [Reichsforstmeister Hermann] Goering's plan began to take shape." Local people also feared "the removal of access to free or cut-price forest goods such as berries, mushrooms, herbs and firewood." Franklin, "Bialowieza Forest," 1480, 1474.

66. Biruta Ēce (Tārgale township mayor), interview by the author, Tārgale, July 21, 1999.

67. Gundars Berkholds (Kolka Livonian Society representative), interview by the author, Kolka, July 23, 1999.

68. Sēle, "Kurzemes jūrmala."

69. Sēle, "Vai jūrmalas balsis."

70. Plaudis, "Šie ciemi dzīvos īpatnēju dzīvi."

71. Egils Kiršpils, "Nacionālais parks—nākotnes perspektīva," Banga, March 27, 1997.

72. Blumberga, "Viens liels dabas parks."

73. Melluma, August 1999 interview.

74. Pilāts, 1999 interview.

75. Laicāns, 1999 interview.

Chapter 6 | Wild Horses in a "European Wilderness"

1. Švābe, *Latvijas vēsture*, 531.

2. World Wide Fund for Nature, Latvian Ministry of Environment, and the World Bank, "Terms of Reference: Management Plans and Programs for Lake Pape and Jūrkalne, Latvia (1994–1997)," October 19, 1994, 2.

3. Phil Macnaghten and John Urry, *Contested Natures* (London: Sage, 1998), 215.

4. World Wide Fund for Nature et al., "Terms of Reference," 3.

5. Identification of Strengths, Weaknesses, Opportunities and Threats pertaining to the planning goal (in Latvian, "*taisīt svotus*").

6. Terms in the glossary included: action, activities, assumptions, chances, external environment, internal situation, implementation, means, objectively verifiable indicator, source of verification, precondition, region, result, risk, reasoning/focus, problem analysis, strategy, strategy analysis, strategic planning, comprehensive planning, target group, development goal—tentative goal—provisional goal—final goal, direct objective —wider objective, optimal scenario—doom scenario—policy scenario—trend scenario, short—medium—long term, participatory approach, top-down—bottom-up approach. Liepājas rajona padome, Liepājas pilsētas dome, Pro-Inter c.v. Flandrija, VARAM, *Liepājas reģiona attīstības un darbības plāns: Projekta pirmās fāzes ziņojums* (unpublished document, 1998), 26–30.

7. Maruta Kaminska (ecosystem inspector, Liepāja regional environmental agency), interview by the author, February 8, 1999.

8. World Wide Fund for Nature et al., "Terms of Reference," 3–4.

9. WWF-Denmark, "Management Plan for Lake Pape Project Area" (unpublished document, 1996), 12.

10. Ibid., 24.

11. Ibid., 4.

12. Ibid., 41.

13. Ibid., 6.

14. Ibid., 5.

15. Valts Vilnītis (former director, MEPRD Department of Environmental Protection), interview by the author, Riga, March 18, 1999.

16. Māris Dadzis (Jūrkalne township mayor), interview by the author, Jūrkalne, February 19, 1999.

17. Nora Driķe, "Ar projektiem dubulto gada budžetu," *Diena*, July 17, 1999.

18. Jānis Veits (Rucava township mayor), address to the conference "Rucava Yesterday, Today and Tomorrow," Rucava, August 6, 1999.

19. Veits, interview by the author, Rucava, February 22, 1999.

20. Driķe, "Ar projektiem," 10.

21. Both Veits and his deputy, interestingly, were land reclamation technicians (*meliorātori*) during the Soviet period. "Land reclamation is close to nature!" Veits declared. "That's why we like the ecological questions!" Veits, 1998 interview.

22. The fund-raising skills acquired by Veits and other local ICZM veterans netted the township a shocking wealth of high-tech resources—several computers, color printers, and four scanners in this tiny village—but not the capacity to make real use of them. Thus the town library had a color printer and state-of-the-art scanner; yet, lacking anything but word-processing software, they stood idle, apt symbols of "visionless" fundraising capacity. For a discussion of how East European NGO agendas are often driven by the availability of Western funding, see Barbara A. Cellarius, "Linking Global Priorities and Local Realities: Nongovernmental Organizations and the Conservation of Nature in Bulgaria," in *Bulgaria in Transition: Environmental Consequences of Political and Economic Transformation*, ed. Krassimira Paskaleva et al. (Aldershot, UK: Ashgate, 1998).

23. Ināra Rūce (chair, Rucava Nature Fund), interview by the author, Rucava, February 22, 1999.

24. Emils Melngailis, *Uz tām prūšu robežām*, quoted in Rucavas Dabas Fonds, *Ir tāda vieta pie Baltijas jūras: Atklāsme par Rucavas pagastu* (Riga: WWF-Latvia and Rucavas Dabas Fonds, ND), 11.

25. Quoted in ibid., 17.

26. Gunta Timbra (founding member, Rucava Nature Fund), interview by the author, Rucava, February 22, 1999.

27. Rucavas Dabas Fonds, *Ir tāda vieta*, 6.

28. Ibid., 13, 17.

29. Gunta Timbra, 1999 interview.

30. Rucavites are not alone, of course, in being viewed by themselves and others as living preservers of ethnographic heritage. Sweden's rural Dalarna region "has become iconic of Swedishness in various ways. . . . Dalarna has been portrayed as one of the places where folk costume persisted into this century, where the modern had not succeeded the folk but coexisted with it. . . . It was atypical of Sweden yet, as an interior that was different and romanticized through the category of folklife, it could be made a national symbol. It became an internal Other, a place against whose fading a modern Sweden could be defined." Mike Crang, "Nation, Region and Homeland: History and Tradition in Dalarna, Sweden," *Ecumene* 6, no. 4 (1999): 458. In Setumaa in southeast Estonia, the Setu people (a distinct ethnic minority like the Livonians in Latvia) have sought to protect their language and cultural distinctiveness. This movement is characterized by "a nostalgic attitude towards local traditional peasant culture and an idealization of the rural

past. . . . Folk costume, folklore and traditional features of the economy are considered especially important." Berg, "Local Resistance," 117.

31. Gedimins Salmiņš (member, Rucava Nature Fund), interview by the author, Rucava, February 22, 1999.

32. "Tos brikšus brist nav baudāms!" Sventāja River Project meeting, Rucava, January 13, 1999.

33. To illustrate the relative significance of distance in this cash-strapped locality, a mail carrier whom I interviewed said she had to cover her 34-kilometer route on a bicycle; in winter she made the trip only every other day.

34. Gunta Timbra, 1999 interview.

35. Voldemārs Timbra (head, Rucava Farmers' Association), interview by the author, Rucava, August 6, 1999.

36. DEVCO Ireland, Piekrastes integrētās pārvaldes un apsaimniekošanas plāns Latvijā un Lietuvā: Ziņojums par pārskatītajiem PIPAP plānu uzmetumiem, sējums IIIe—Nīca/Rucava (unpublished document, 1997), 25–28.

37. IUCN, Parks for Life: Report of the IVth World Congress on National Parks and Protected Areas (Gland, Switzerland: IUCN, 1993), 40.

38. Ilbery and Bowler, "From Agricultural Productivism," 102.

39. Brian Gardner, European Agriculture in the New Millenium, vol. 2 (London: Agra Europe, 1999), 254.

40. The first American restoration ecology project is considered to be that of Aldo Leopold and his colleagues at the University of Wisconsin's arboretum in Madison, where Curtis Prairie was restored to replicate the structure, function, and dynamics of historic plant communities.

41. Jan Pouwel Bakker, Nature Management by Grazing and Cutting: On the Ecological Significance of Grazing and Cutting Regimes Applied to Restore Former Species-rich Grassland Communities in the Netherlands (Dordrecht: Kluwer Academic Publishers, 1989), 10.

42. Ibid.; Frans W. M. Vera, "Metaphors for the Wilderness: Oak, Hazel, Cattle and Horse" (Ph.D. diss., Agricultural University, Wageningen, The Netherlands, 1997); Michiel F. Wallis De Vries, "Large Herbivores and the Design of Large-Scale Nature Reserves in Western Europe," Conservation Biology 9, no. 1 (1995): 25–33; and S. E. Van Wieren, "The Potential Role of Large Herbivores in Nature Conservation and Extensive Land Use in Europe," Biological Journal of the Linnaean Society 56 (Supplement, 1995): 11–23.

43. Vera, "Metaphors for the Wilderness," Epilogue.

44. Wallis De Vries, "Large Herbivores," 31.

45. Nature development efforts have accelerated since the passage in 1990 of an ambitious National Nature Policy Plan aimed not only at conserving existing nature areas and establishing connecting corridors between them, but also at creating new nature areas on a dramatic scale. The plan calls for taking 150,000 hectares, or 7.5% of the country's total agricultural area, out of production over a thirty-year period. "This area is, at the moment, partly wasteland from an agricultural productivity point of view. Other parts are good, or even highly productive, agricultural land." D. Strijker, F. J. Sijtsma, and D. Wiersma, "Evaluation of Nature Conservation: An application to the Dutch ecological network," *Environmental and Resource Economics* 16, no. 4 (2000): 364.

46. Large herbivores and carnivores are viewed as useful targets for focusing biodiversity protection campaigns for a number of reasons. Ensuring sufficient numbers of large mammals at the top of the food chain enables ecosystems to function properly and reproduce themselves, and large herbivores in particular are vital "keystone species" because of their role in maintaining landscape and vegetative diversity. Moreover, as wide-ranging species requiring large living areas, large mammals serve as "umbrella species": by protecting their habitats, many other plant and animal habitats are protected as well. Finally, as "charismatic megafauna" that people enjoy seeing in the wild, large mammals help popularize biodiversity protection efforts through their tourism value. See www.largecarnivores-lcei.org.

47. From the WWF-International website, www.panda.org.

48. This movement, sometimes called "ethical" restorationism, is only one strand in the complex of U.S. restoration practices. Utilitarian restorationism, by contrast, focuses on restoring ecosystem functions that are "useful" to humans: recreating wetlands, for example, to replace lost functions such as flood control or watershed protection.

49. Frederick Turner, "A Field Guide to the Synthetic Landscape: Toward a New Environmental Ethic," *Harper's* 276, no. 4 (1988): 50. See also Turner, "Cultivating the American Garden: Toward a Secular View of Nature," *Harper's* 271, no. 8 (1985): 45–52.

50. William R. Jordan III, "'Sunflower Forest': Ecological Restoration as the Basis for a New Environmental Paradigm," in *Beyond Preservation: Restoring and Inventing Landscapes*, ed. A. Dwight Baldwin Jr., Judith De Luce, and Carl Pletsch (Minneapolis: University of Minnesota Press, 1994), 19.

51. Ibid., 21.

52. Olwig, "Reinventing Common Nature," in Cronon, *Uncommon Ground*, 396.

53. John Blunden and Graham Turner, *Critical Countryside* (London: British Broadcasting Company, 1985), 150.

54. Ibid., 44.

55. Lowenthal, "European and English Landscapes," 29.

56. Laura L. Jackson, "The Role of Ecological Restoration in Conservation Biology," in Fiedler, *Conservation Biology*, 435.

57. While the Dutch have been its leading proponents, antiagrarian restoration is embraced by other European conservationists as well. Thus, for example, R. A. Jarman of the United Kingdom's National Trust has "argued for some time against conservationists' fixation with mimicking 'traditional' management and [has] pressed, for example, for ancient woodlands to be restored to 'wildwood.'" Jarman, "Ecological Restoration: the End of *Status* quo-ism in the National Trust?" *Biological Journal of the Linnean Society* 56 (Supplement, 1995): 213.

58. Willem Overmars et al., *Lake Pape: Restoring a European Wilderness. Perception of Nature and Employment Opportunity* (unpublished document, 1998), 7.

59. Ibid., 7.

60. These traits include "the ability to . . . foal without help, and at an early age; resistance against cold, snow and rain . . . ; the ability to get nutrients from withered grass, or leaves, twigs and barks; a limited milk supply with high nutritional value in combination with a small udder; a physique that is adequate for the wild terrain; alert and confident behaviour; well-developed motherly instincts; the ability to develop a layer of subcutaneous fat and between the muscles in the growing season to get them through winter without having to draw on muscle tissues." Ibid., 51.

61. Overmars, conversation with the author, Rucava, November 20, 1998.

62. Overmars et al., *Lake Pape*, 47.

63. While portions of the territory had been under various protection regimes since the Soviet period, there was no coordination among regimes and no common management rules and practices.

64. Overmars et al., *Lake Pape*, 7, 29.

65. Ibid., 30.

66. Ibid., 57–58.

67. Uģis Rotbergs (director, WWF-Latvia), interview by the author, Riga, July 9, 1999.

68. Ibid.

69. Jānis Priednieks (chairman, Latvian Fund for Nature), interview by the author, Riga, July 9, 1999. Interestingly, in 2002 the LDF introduced five Koniks, donated by the same Dutch consultancy (Stifting Ark) that provided the Lake Pape horses, into one of its own project territories, the Lake Engure Nature Park in a coastal wetland on the Gulf of Riga. But the authors of the LDF's Interim Activity Report for the project express similar reservations about management for natural processes, noting that introducing the Koniks was "not foreseen in the project" and that: "There is an argument between nature conservationists, supporting natural grazing and semi-natural grazing approaches. In our project we are following the principles of semi-natural grazing, as we think that it is more beneficial for local population. But we also want to keep the other options open and test the approach of natural grazing in Engure. Probably, the mixture of both ap-

proaches will be the best solution for Engure." Inga Račinska, "Implementation of Management Plan for the Lake Engure Nature Park" (unpublished document, September 15, 2002).

70. "'There is a kind of irrational xenophobia about invading animals and plants that resembles the inherent fear and intolerance of foreign races, cultures, and relgions,' one biologist . . . told a recent scientific conference devoted to exotics." Budiansky, *Nature's Keepers*, 130.

71. On local resistance to introducing wild horse in the United States, see J. Sanford Rikoon, "Wild Horses and the Political Ecology of Nature Restoration in the Missouri Ozarks," presented at the conference "Political Ecology at Home," Rutgers University Department of Geography, March 2003. The politics of mammal reintroduction tends to be much more heated when carnivores are involved; see Alec Brownlow, "A Wolf in the Garden: Ideology and Change in the Adirondack Landscape," in *Animal Spaces, Beastly Places: New Geographies of Human-Animal Relations*, ed. Chris Philo and Chris Wilbert (London: Routledge, 2000); and Matthew A. Wilson, "The Wolf in Yellowstone: Science, Symbol or Politics? Deconstructing the Conflict between Environmentalism and Wise Use," *Society & Natural Resources* 10, no. 5 (1997): 453–68.

72. Rucavas Dabas Fonds, *Ir tāda vieta*, 12.

73. Ibid., 7.

74. Ethnic Latvians comprise 93% of the population of Rucava township.

75. See Marta Bruno, "Playing the Co-operation Game: Strategies around International Aid in Post-socialist Russia," in *Surviving Post-Socialism: Local Strategies and Regional Responses in Eastern Europe and the Former Soviet Union*, ed. Sue Bridger and Frances Pine (London and New York: Routledge, 1998); Sampson, "The Social Life of Projects"; and Cellarius, "Linking Global Priorities."

76. Brian Slocock, "'Whatever Happened to the Environment?' Environmental Issues in the Eastern Enlargement of the European Union," in *Back to Europe: Central and Eastern Europe and the European Union*, ed. Karen Henderson (London: UCL Press, 1999), 156.

77. Dadzis, 1999 interview.

78. See Cellarius, "Linking Global Priorities," on the difficulties national environmental NGOs in Eastern Europe often face in establishing meaningful collaboration with local organizations and individuals.

Chapter 7 | "Lichens Are Not Our National Treasure"

1. Nancy Langston, "Forest Dreams, Forest Nightmares: An Environmental History of a Forest Health Crisis," in *American Forests: Nature, Culture, and Politics*, ed. Char Miller (Lawrence: University Press of Kansas, 1997), 249, 258.

2. Ibid., 260.

3. Phil McManus, "Histories of Forestry: Ideas, Networks and Silences," *Environment and History* 5, no. 2 (1999): 192.

4. Joachim Radkau, "Wood and Forestry in German History: In Quest of an Environmental Approach, *Environment and History* 2 (1996): 63–76; quoted in ibid.

5. McManus, "Histories of Forestry," 191.

6. Aldo Leopold, "Wilderness," in *The Great New Wilderness Debate*, ed. J. Baird Callicott and Michael Nelson (Athens: University of Georgia Press, 1998), 518–19.

7. J. Matīss, "Mežu ierīcības attīstība Latvijas PSR," *Mežsaimniecība un Mežrūpniecība*, 1977, no. 3: 13.

8. Heinrihs Strods, "Latvijas mežu politika un likumdošana (XI gs.–1940.g.)," in *Latvijas mežu vēsture līdz 1940. gadam*, ed. Strods (Riga: WWF, 1999), 102.

9. Anon., "Darbu sākot," *Meža Dzīve*, 1925, no. 1: 2.

10. "II. virsmežžiņu un mežžiņu kongresa stenograma," February 20, 1922, 21.

11. K. Z., "Turēsim augsti savu prestižu," *Meža Dzīve*, 1929, no. 42: 1329.

12. "Šalkonis," "Mežu darbinieki latvju daiņās," *Meža Dzīve*, 1926, no. 16: 497.

13. To be precise, responsibility for forest management and logging were divided between separate ministries of forestry and forest industry from 1947 to 1957, when the two ministries were merged.

14. K. Bušs, "Meža tipoloģija," in *Latvijas Meži*, ed. M. Bušs and J. Vanags (Riga: Avots, 1987), 72.

15. Central Statistical Bureau, *Latvia: Chief Statistical Indicators 1999*, 26.

16. "Latvijas nafta ir kokmateriali un mezs," *Diena*, January 10, 1997.

17. Rotbergs, "Kādiem jābūt mežiem Latvijā?" *Diena*, December 1, 1992.

18. For an overview of the New Forestry and other new approaches to biodiversity conservation in forests, see William S. Alverson, Walter Kuhlmann, and Donald M. Walker, *Wild Forests: Conservation Biology and Public Policy* (Washington, DC: Island Press, 1994).

19. Osvalds Cinītis (director, MEPRD Environmental Inspection Flora and Fauna Unit), interview by the author, Riga, February 12, 1996. Brigita Laime (biologist, University of Latvia and Latvian Fund for Nature), interview by the author, Riga, February 13, 1996.

20. Jānis Jurģis and Uldis Kalnietis (senior specialists, MEPRD Environmental Protection Division), interview by the author, Riga, February 21, 1996. Many protected categories lack even a presumptive connection to ecology: protected corridors along roadsides, for example, were established by the Russian and Soviet powers to guarantee readily accessible timber reserves in times of military emergency. After devastating

windfalls in the late 1960s, large swaths of forest were indiscriminately placed under protection to reduce total harvesting and to replenish Latvia's forest endowment.

21. Jānis Priednieks, Māris Kreilis, and Ilona Lodziņa, *National Biodiversity Action Plan for Latvia* (Riga: MEPRD, 1995), 12.

22. Jānis Priednieks (chairman, Latvian Fund for Nature), interview by the author, Riga, February 19, 1996.

23. Arnis Melnis (director, State Forest Service), interview by the author, Riga, July 30, 1999. Jānis Rubenis (former deputy minister of forestry), interview by the author, Babīte, August 10, 1999.

24. Uldis Grava (director, SFS Logging Unit), interview by the author, Riga, February 8, 1996.

25. To understand the counterhegemonic significance of controlled burns in Latvia, consider the view of forest fires held by great tree liberator Guntis Eniņš: "Behind the next hill the painting suddenly crumbles and the view suffocates in a terrible landscape. The harmony of nature's beauty has been taken over by the horror of the forest—fire." For Eniņš, man's role is to protect nature from itself: "Who will stop the fire? Who will save the forest from burning down altogether? Only man. Only man can stand against the fire-plague, the fire disaster." Eniņš, *Tepat Latvijā*, 64–66.

26. Normunds Priedītis, *Latvian Forests: Nature and Diversity* (Riga: WWF, 1999).

27. Magnus Sylven, "WWF in Latvia: Status and Future Opportunities" (unpublished document, 1998), 5.

28. Uvis Suško et al., *Natural Forests of Latvia: A study on biodiversity structures, dependent species and forest history* (Riga: WWF, 1997).

29. Katrina Z. S. Schwartz, *The Politics of Sustainable Forestry in Latvia: Property, Enterprise and the State in Transition from Communism* (Riga: WWF, 1996). I authored this study, based on forty open-ended interviews conducted in the winter and spring of 1996, as an independent consultant for WWF.

30. Manidis Roberts International, *WWF Project LV0005—Latvia: Privatisation and Sustainable Forestry. Mid-Term Evaluation Report* (unpublished document, 1997), 18.

31. Ēvalds Pozņakovs (executive director, Rēzeknes MRS), interview by the author, Riga, March 19, 1999.

32. Arnis Melnis (director, State Forest Service, and former chief engineer, SIA Silva), interview by the author, Riga, March 29, 1999.

33. Kaspars Liepiņš (Mežole project assistant), interview by the author, Smiltene, March 31, 1999.

34. Guntars Cīrulis (Mežole district state forester), interview by the author, Smiltene, April 1, 1999.

35. Manidis Roberts International, *WWF Project LV0005*, 7.

36. Imants Mangalis (professor, Agricultural University of Latvia Department of Forestry), interview by the author, Jelgava, February 5, 1996.

37. In 1995, stumpage fees and fines accounted for 62% of the 11.4 million-lat forestry budget. Schwartz, *Politics of Sustainable Forestry*, 11.

38. Gunārs Dišlers (Ogre district chief forester), interview by the author, Ogre, December 7, 1995.

39. Arnis Melnis (chief engineer, SIA Silva), interview by the author, Riga, December 4, 1995.

40. Manidis Roberts, *WWF Project LV0005*, 12.

41. According to the review, the most fundamental weaknesses of the Latvian system were state control of all four functions of forestry governance—the regulatory, oversight, ownership and production functions—and the conflicts of interest arising from the fiscal dependence of the first two functions on the second two. See State Forest Service, *Latvia Forestry Sector Master Plan: Summary Final Report* (unpublished document, 1995), 17–23. See also Schwartz, *Politics of Sustainable Forestry*, for a detailed discussion of the review's recommendations.

42. Rotbergs, "Kā mans dēls Robītis novērtēs mūsu izaudzēto mežu?" *Meža Dzīve*, 1992, no. 5: 22–23.

43. World Wide Fund for Nature, *Privatisation and Sustainable Forestry in Latvia: Mežole Demonstration Project. Final Report* (unpublished document, 1999), 15.

44. Sylven, "WWF in Latvia," 5.

45. Leons Vītols (director, SIA Latvijas Mežs, and former minister of forestry), interview by the author, Riga, June 14, 1999.

46. Pēteris Zālītis (senior scientist, Silava Forest Science Institute), interview by the author, Salaspils, July 7, 1999.

47. Zālītis, "Piepes nav mūsu valsts lielākā bagātība," *Meža Dzīve* 1997, 12: 16–17.

48. Ibid.

49. Mārtiņš Dāboliņš, "Bez laba Meža likuma nebūs kārtīga Latvijas meža," *Latvijas Vēstnesis*, November 21, 1997.

50. K. Ērglis, "Mežs un pārpurvošanās," *Meža Dzīve*, 1994, no. 4: 7.

51. Māris Daugavietis, "Ar skatu no malas par ilgtspējīgu mežsaimniecību un bioloģisko daudzveidību (neprofesionāļa viedoklis)," *Meža Dzīve*, 1998, no. 11/12: 6.

52. Gunārs Dišlers (Ogre district chief forester), interview by the author, Ogre, December 7, 1995.

53. Aija Zviedre, "Ja mežīpašnieks dzīvo pilsētā," *Meža Dzīve*, 1997, no. 9: 4.

54. Jānis Vazdiķis (director, Latvian State Forest Inventory Institute), interview by the author, Salaspils, December 14, 1995.

55. Juris Matīss (former director, Latvian State Forest Inventory Institute), interview by the author, Salaspils, June 29, 1999.

56. Lāsma Āboliņa (senior official for environmental protection and forest monitoring, SFS Forests Department Forest Inventory and Environmental Protection Unit), interview by the author, Riga, November 23, 1995.

57. Imants Ansons (director, Ogres MRS), interview by the author, Riga, December 6, 1995.

58. Zalītis, "Meža ekoloģija un mežsaimniecība," *Mežsaimniecība un Mežrūpniecība*, 1990, no. 1: 11.

59. Zālītis, "Par mežzinātni Latvijā vakar un šodien," *Meža Dzīve*, 1996, no. 7: 23.

60. Zalītis, "Pārmitrie meži, to meliorācija un dabas daudzveidības saglabāšana Latvijā," *Meža Dzīve*, 1995, no. 9: 8–13.

61. Ibid., 8; Zālītis, 1999 interview; "Piepes nav," 16.

62. M. Bušs, "Šalkones vecbiedra M. Buša referāts," *Meža Dzīve*, 1993, no. 8: 21.

63. Zālītis, "Par mežzinātni Latvijā," 24.

64. Aija Zviedre (president, Latvian Forest Employees' Association), interview by the author, Riga, March 23, 1999.

65. Zviedre is playing on the fact that *ozols* means oak in Latvian. Trees provide many other very common Latvian surnames: Alksnis (alder), Apse (aspen), Kļava (maple), Lazda (hazelnut), Liepa (linden), Osis (ash), Paeglis (juniper), Vītols (willow), along with diminutives such as Apsītis, Bērziņš (birch), Eglītis (spruce), Gobiņš (elm), Priedītis (pine), and Vīksniņš (elm).

66. Zviedre, "Mežsargs manā atmiņā," *Mežsaimniecība un Mežrūpniecība*, 1991, no. 3: 42–43.

67. M. Bušs, "Baltijas valstu Mežkopju savienības izveidošanās," *Mežsaimniecība un Mežrūpniecība*, 1991, no. 4: 11.

68. Zālītis, "Mežkopība, mežsaimniecība, mežzinātne," *Meža Dzīve*, 1997, no. 8: 8.

69. Dāboliņš, "Bez laba Meža likuma," 5.

70. Daugavietis, "Ar skatu no malas," 11.

71. "Latvijas meža darbinieku biedrības 5. kongresa rezolūcija," *Meža Dzīve*, 1997, no. 5: 36.

72. Jurģis Jansons, "Bioloģiskā daudzveidība Latvijā—vai diskusijai ir robežas?" *Meža Dzīve*, 1998, no. 7: 7.

73. Georgs Gavrilovs (director, SFS Forests Department), interview by the author, Riga, November 23, 1995.

74. Jānis Osis (director, SFS Forest Inventory and Environmental Protection Unit), interview by the author, Riga, November 23, 1995.

75. Jaunsudrabiņš, *Jaunsaimnieks un velns*, 34.

76. Quoted in Aizsilnieks, *Latvijas Saimniecības Vēsture*, 609.

77. Ibid., 241.

78. Quoted in ibid., 721.

79. Quoted in ibid., 607.

80. For example, transportation was subsidized so that all farmers could sell their produce at the same prices, regardless of distance from markets, thereby disabling geographical incentives to specialize, and anti-automobile policies promoted inefficient, labor-intensive milk deliveries by horse from one or two farms at a time. Ibid., 718, 746.

81. Ibid., 830–31. See also Nicholas W. Balabkins, "Latvia's Economic Nationalism," *East European Quarterly* 16, no. 2 (1982): 151–69.

82. Andis Krēsliņš (Alūksne district chief forester), interview by the author, Alūksne, May 30, 1999.

83. Zviedre, interview by the author, Riga, January 30, 1996.

84. M. Bušs, quoted in I. Mangalis, "Latvijas mežu apsaimniekošanas problēmas un uzdevumi," *Mežsaimniecība un Mežrūpniecība*, 1990, no. 1: 18.

85. Viesturs Lārmanis, "Kas ir un kas nav bioloģiski pilnvērtīgs mežs?" *Meža Dzīve*, 1998, no. 1: 16.

86. In each participating country, the FSC approves locally drafted standards for ascertaining the sustainability of silvicultural and logging practices and accredits local bodies to certify compliance with these standards.

87. Guntars Lagūns (coordinator, FSC task force), interview by the author, Riga, March 25, 1999.

88. Roberts Strīpnieks (director, FAO-SFS project for Strengthening the Institutional Capacity of the Latvian Forest Authority), interview by the author, Riga, July 12, 1999.

89. Priednieks, 1999 interview.

90. Ģirts Strazdiņš (president, Environmental Protection Club of Latvia), interview by the author, Riga, July 30, 1999.

91. V. Eiche, "Cilvēka loma Latvijas meža izveidošanas gaitā," in *Mežsaimniecības rakstu krājums, XV sējums* (Riga: Latvijas mežkopju un meža darbinieku biedrība, 1937), 134–35; Krišjānis Lange, "Par meža tīrīšanu," in *Mežsaimniecības rakstu krājums, I sējums* (Riga: Latvijas mežkopju un meža darbinieku biedrība, 1923), 79; H. Upītis, "Modernā mežsaimniecība pēc A. Kubelkas," in *Mežsaimniecības rakstu krājums, I sējums* (Riga: Latvijas mežkopju un meža darbinieku biedrība, 1923), 62; H. Upītis, *Rokas grāmata mežkopjiem*, vol. 4: *Mežkopība* (Rīga: Mežu departaments, 1939), 164; K. Kiršteins, *Rokas grāmata mežkopjiem*, vol. 3: *Mežzinība* (Rīga: Mežu departaments, 1926), 43–47; K. Melderis, *Mācība par mežu* (Riga: Valters un Rapa, 1939), 114; E. Kalniņš, *Mežu kopšanas cirtes* (Riga: Mežu departaments, 1938), 9–10, 17–19.

92. Rotbergs, interview by the author, Riga, July 9, 1999.

93. Strods, *Latvijas mežu vesture*.

94. Normunds Priedītis, interview by the author, Riga, August 12, 1999.

95. Krišjānis Valdemārs, "Vai dabas zinātnības ir vajadzīgas latviešu lasītāju lielākai daļai?" in *Tēvzemei*, 219–24.

96. Zālītis, "Klasiskās vērtības un īslaicīgas parādības Latvijas mežsaimniecībā," *Meža Dzīve*, 1998, no. 5: 5.

97. Jansons, "Bioloģiskā daudzveidība Latvijā—problēma vai ambīcijas?" *Meža Dzīve*, 1998, no. 3: 9.

98. Priedītis, "Bioloģiskā daudzveidība Latvijā—vai dogmām eksistē robežas?" *Meža Dzīve*, 1998, no. 6: 12.

99. Jansons, "Bioloģiskā daudzveidība Latvijā—vai diskusijai ir robežas?" *Meža Dzīve*, 1998, no. 7: 6, emphasis in original.

100. Priedītis, "Bioloģiskā daudzveidība Latvijā," 12.

101. Zālītis, "Piepes nav," 16.

Conclusion

1. http://www.panda.org/about_wwf/where_we_work/europe/what_we_do/policy_and_events/epo/news.

2. Yi-Fu Tuan, "Perceptual and Cultural Geography: A Commentary," *Annals of the American Association of Geographers* 93, no. 4 (2003): 880.

3. Lowenthal, "European and English Landscapes," 29.

4. Tuan, "Perceptual and Cultural Geography," 880.

5. Hobsbawm, *Nations and Nationalism Since 1780*, 169.

6. Ernest Gellner concurs: "Modern nationalism . . . wells up in otherwise anonymous, atomized populations. . . . The destruction of civil society in the past seventy years has

prepared the soil for the rapid sprouting of nationalism." "Nationalism in the Vacuum," 250–53.

7. Hobsbawm, *Nations and Nationalism Since 1780*, 170, 127, 176, 191, 186.

8. Roman Szporluk, *Communism and Nationalism: Karl Marx versus Friedrich List* (Oxford: Oxford University Press, 1988), 10.

9. A. Smith, *Ethnic Origins of Nations*, 208, 177.

10. Mary Kaldor, "Cosmopolitanism Versus Nationalism: The New Divide?" in *Europe's New Nationalism: States and Minorities in Conflict*, ed. Richard Caplan and John Feffer (New York: Oxford University Press, 1996), 43.

11. David Morley and Kevin Robins, "No Place Like *Heimat*: Images of Home(land) in European Culture," in *Space and Place: Theories of Identity and Location*, ed. Erica Carter, James Donald, and Judith Squires (London: Lawrence & Wishart, 1993), 26–27.

12. "Indeed, 'homeland' territoriality remains a fundamental organizing principle of modern Europe," argue Zsuzsa Csergo and James M. Goldgeier, in "Nationalist Strategies and European Integration," *Perspectives on Politics* 2, no. 1 (2004): 32. On the continued salience of place-based national identities, see also Craig Young and Duncan Light, "Place, National Identity and Post-Socialist Transformations: An Introduction," *Political Geography* 20 (2001): 941–55.

13. Kaldor, "Cosmopolitanism Versus Nationalism" (citing Robert Reich), 47.

14. Appadurai, *Modernity at Large*, 4.

15. Verdery, "Nationalism, Postsocialism, and Space in Eastern Europe," *Social Research* 63, no. 1 (1996): 90–91. Catherine Wanner agrees: "The nineteenth-century peasant, billed as the national essence, is strangely out of place . . . in a highly educated and industrialized country." *Burden of Dreams*, 93.

BIBLIOGRAPHY

Aasland, Aadne, and Tone Flotten. "Ethicity and Social Exclusion in Estonia and Latvia." *Europe-Asia Studies* 53, no. 7 (2001): 1023–49.

Abrahams, Ray, ed. *After Socialism: Land Reform and Social Change in Eastern Europe*. Providence, RI: Berghahn Books, 1996.

Adams, Jonathan S., and Thomas O. McShane. *The Myth of Wild Africa: Conservation Without Illusion*. Berkeley and Los Angeles: University of California Press, 1992.

Aizsilnieks, Arnolds. *Latvijas Saimniecības Vēsture*. Sweden: Daugava, 1968.

Akurāters, Jānis. *Kalpa zēna vasara: atmiņas zīmējums* Riga: Latvijas valsts izdevniecība, 1956.

Alanen, Ilkka, Jouko Nikula, Helvi Poder, and Rein Ruutsoo, eds. *Decollectivization, Destruction and Disillusionment: A Community Study in Southern Estonia*. Aldershot, UK: Ashgate, 2001.

Alverson, William S., Walter Kuhlmann, and Donald M. Walker. *Wild Forests: Conservation Biology and Public Policy*. Washington, DC: Island Press, 1994.

Anderson, Benedict. *Imagined Communities: Reflections on the Origin and Spread of Nationalism*. London: Verso, 1983.

Anderson, Mikael Skou. "Ecological Modernization or Subversion? The Effect of Europeanization on Eastern Europe." *American Behavioral Scientist* 45, no. 9 (2002): 1394–1416.

Appadurai, Arjun. *Modernity at Large: Cultural Dimensions of Globalization*. Minneapolis: University of Minnessota Press, 1996.

Applebaum, Anne. *Between East and West: Across the Borderlands of Europe*. New York: Random House, 1994.

Applegate, Celia. *A Nation of Provincials: The German Idea of Heimat*. Berkeley: University of California Press, 1990.

Armstrong, John A. *Nations before Nationalism*. Chapel Hill: University of North Carolina Press, 1982.

Arnswald, Sven. *EU Enlargement and the Baltic States: The Incremental Making of New Members*. Helsinki and Bonn: Ulkopoliittinen instituutti and Institut für Europäische Politik, 2000.

Baker, Susan, Maria Kousis, Dick Richardson, and Stephen Young. "Introduction: The Theory and Practice of Sustainable Development in EU Perspective." In Baker et al., *The Politics of Sustainable Development*.

Baker, Susan, Maria Kousis, Dick Richardson, and Stephen Young, eds. *The Politics of Sustainable Development: Theory, Policy and Practice within the European Union*. London: Routledge, 1997.

Bakker, Jan Pouwel. *Nature Management by Grazing and Cutting: On the Ecological Significance of Grazing and Cutting Regimes Applied to Restore Former Species-rich Grassland Communities in the Netherlands*. Dordrecht: Kluwer Academic Publishers, 1989.

Balabkins, Nicholas. "Latvia's Economic Nationalism, 1934–1940." *East European Quarterly* 16, no. 2 (1982): 151–69.

Banac, Ivo, and Katherine Verdery, eds. *National Character and National Ideology in Interwar Eastern Europe*. New Haven: Yale Center for International and Area Studies, 1995.

Bardenstein, Carol B. "Trees, Forests, and the Shaping of Palestinian and Israeli Collective Memory." In *Acts of Memory: Cultural Recall in the Present*, edited by Mieke Bal, Jonathan Crewe, and Leo Spitzer. Hanover, NH: University Press of New England, 1999.

Barnard, F. M. *Herder's Social and Political Thought: From Enlightenment to Nationalism*. Oxford: Clarendon Press, 1965.

Bassin, Mark. "Russian Geographers and the 'National Mission' in the Far East." In Hooson, *Geography and National Identity*.

———. "Turner, Solov'ev, and the 'Frontier Hypothesis': The Nationalist Signification of Open Spaces." *Journal of Modern History* 65 (September 1993): 473–511.

Beaufoy, Guy, David Baldock, and Julian Clark. *The Nature of Farming: Low Intensity Farming Systems in Nine European Countries*. London: Institute for European Environmental Policy, 1994.

Beissinger, Mark. *Nationalist Mobilization and the Collapse of the Soviet State: A Tidal Approach to the Study of Nationalism*. New York: Cambridge University Press, 2001.

Beitnere, Dagmāra. "Vai latvieši ir tikai zemnieki?" *Karogs* 11 (2002): 150–63.

Berg, Eiki. "Local Resistance, National Identity and Global Swings in Post-Soviet Estonia." *Europe-Asia Studies* 54, no. 1 (2002): 109–22.

Berg, Eiki, and Saima Oras. "Writing Post-Soviet Estonia on to the World Map." *Political Geography* 19 (2002): 601–25.

Biehl, Janet, and Peter Staudenmaier. *Ecofascism: Lessons from the German Experience*. Edinburgh and San Francisco: AK Press, 1995.

Billig, Michael. *Banal Nationalism*. London: Sage, 1994.

Blaumanis, Rūdolfs. *Indrāni*. Rockville, MD: ALA Kultūras birojs, 1965.

Blunden, John, and Graham Turner. *Critical Countryside*. London: British Broadcasting Company, 1985.

Boruks, Artūrs. *Zemnieks, zeme un zemkopība Latvijā: no senākiem laikiem lidz mūsdienām*. Riga: Grāmatvedis, 1995.

Bradshaw, Michael, and Alison Stenning, eds. *East Central Europe and the Former Soviet Union: the Post-Socialist States*. Harlow, UK: Pearson Education, 2004.

Bramwell, Anna. *Ecology in the Twentieth Century*. New Haven: Yale University Press, 1989.

Brownlow, Alec. "A Wolf in the Garden: Ideology and Change in the Adirondack Landscape." In *Animal Spaces, Beastly Places: New Geographies of Human-Animal Relations*, edited by Chris Philo and Chris Wilbert. London: Routledge, 2000.

Brubaker, Rogers. *Citizenship and Nationhood in France and Germany.* Cambridge: Harvard University Press, 1992.

———. *Nationalism Reframed: Nationhood and the National Question in the New Europe.* Cambridge: Cambridge University Press, 1996.

Bryant, Raymond L. "Power, Knowledge and Political Ecology in the Third World: A Review." *Progress in Physical Geography* 22, no. 1 (1998): 79–94.

Bryant, Raymond L., and Sinead Bailey. *Third World Political Ecology.* London: Routledge, 1997.

Bryden, John. *Towards an Agenda for Sustainable Rural Development in Europe: A Discussion Paper.* Aberdeen, UK: Arkleton Centre for Rural Development Research, 1998.

Bruno, Marta. "Playing the Co-operation Game: Strategies around International Aid in Post-Socialist Russia." In *Surviving Post-Socialism: Local Strategies and Regional Responses in Eastern Europe and the Former Soviet Union,* edited by Sue Bridger and Frances Pine. London and New York: Routledge, 1998.

Budiansky, Stephen. *Nature's Keepers: The New Science of Nature Management.* New York: Free Press, 1995.

Buller, Henry, Geoff A. Wilson, and Andreas Holl. *Agri-Environmental Policy in the European Union.* Aldershot, UK: Ashgate Publishing, 2000.

Bungs, Dzintra. *The Baltic States: Problems and Prospects of Membership in the European Union.* Baden-Baden: Nomos Verlagsgesellschaft, 1998.

Bunkše, Edmunds V. "The Case of the Missing Sublime in Latvian Landscape Aesthetics and Ethics." *Ethics, Place and Environment* 4, no. 3 (2001): 235–46.

———. "God, Thine Earth Is Burning: Nature Attitudes and the Latvian Drive for Independence." *GeoJournal* 26, no. 2 (1992): 203–09.

———. "Reality of Rural Landscape Symbolism in the Formation of a Post-Soviet, Postmodern Latvian Identity." *Norsk geografisk Tidsskrift* 53 (1999): 121–38.

Burnham, Philip. *Indian Country, God's Country: Native Americans and the National Parks.* Washington, DC: Island Press, 2000.

Bušs, K. "Meža tipoloģija." In *Latvijas Meži,* edited by M. Bušs and J. Vanags. Riga: Avots, 1987.

Buttimer, Anne. "Edgar Kant and Balto-Skandia: *Heimatkunde* and Regional Identity." In Hooson, *Geography and National Identity.*

Čakars, O., A. Grigulis, and M. Losberga. *Latviešu literatūras vēsture no pirmsākumiem līdz XIX gadsimta 80. gadiem.* Riga: Zvaigzne, 1990.

Carmin, JoAnn, and Stacy D. VanDeveer, eds. *EU Enlargement and the Environment: Institutional Change and Environmental Policy in Central and Eastern Europe.* London: Routledge, 2005.

Carruthers, Jane. "Creating a National Park, 1910 to 1926." *Journal of Southern African Studies* 15, no. 2 (1989): 188–216.

Cellarius, Barbara A. "Linking Global Priorities and Local Realities: Nongovernmental Organizations and the Conservation of Nature in Bulgaria." In *Bulgaria in Transition: Environmental Consequences of Political and Economic Transformation,* edited by Kassimira Paskaleva, Philip Shapira, John Pickles, and Boian Koulov. Aldershot, UK: Ashgate, 1998.

Central Statistical Bureau. *Latvia: Chief Statistical Indicators 1999*. Riga: Central Statistical Bureau, 1999.

———. *Statistical Yearbook of Latvia 1998*. Riga: Central Statistical Bureau, 1998.

Chinn, Jeff, and Robert J. Kaiser. *Russians as the New Minority: Ethnicity and Nationalism in the Soviet Successor States*. Boulder: Westview Press, 1996.

Chinn, Jeff, and Lise A. Truex. "The Question of Citizenship in the Baltics." *Journal of Democracy* 7, no. 1 (1996): 132–47.

Cimdiņa, Ausma. "'Sava kaktiņa, sava stūrīša zemes' sakāma vārda tiesības, jeb kāpēc šo stāstu neizlasīt līdz galam?" *Karogs* 12 (1998): 118–28.

Clout, Hugh. "The European Countryside: Contested Space." In *Modern Europe: Place, Culture and Identity*, edited by Brian J. Graham. London: Arnold, 1998.

Cochrane, Nancy J. "Farm Restructuring in Central and Eastern Europe." *Soviet and Post-Soviet Review* 21, nos. 2–3 (1994): 319–35.

Conquest, Robert. *The Harvest of Sorrow: Soviet Collectivization and the Terror-Famine*. New York: Oxford University Press, 1986.

Crang, Mike. "Nation, Region and Homeland: History and Tradition in Dalarna, Sweden." *Ecumene* 6, no. 4 (1999): 447–70.

Creed, Gerald W. "The Politics of Agriculture: Identity and Socialist Sentiment in Bulgaria." *Slavic Review* 54, no. 4 (Winter 1995): 843–68.

Cronon, William, ed. "Introduction: In Search of Nature." In Cronon, *Uncommon Ground*.

———. "The Trouble with Wilderness; or, Getting Back to the Wrong Nature." In Cronon, *Uncommon Ground*.

———. *Uncommon Ground: Rethinking the Human Place in Nature*. New York: W. W. Norton, 1996.

Csaki, Csaba, and Zvi Lerman. "Land Reform and Farm Restructuring in East Central Europe and CIS in the 1990s: Expectations and Achievements after the First Five Years." *European Review of Agricultural Economics* 24 (1997): 428–52.

Csergo, Zsuzsa, and James M. Goldgeier. "Nationalist Strategies and European Integration." *Perspectives on Politics* 2, no. 1 (2004): 21–37.

Darier, Éric. "Foucault and the Environment: An Introduction." In *Discourses of the Environment*, edited by Éric Darier. Oxford: Blackwell, 1999.

Darst, Robert G., Jr. "Bribery and Blackmail in East-West Environmental Politics." *Post-Soviet Affairs* 13, no. 1 (1997): 42–77.

———. "Environmentalism in the USSR: The Opposition to the River Diversion Projects." *Soviet Economy* 4, no. 3 (1988): 223–52.

Dawson, Jane I. *Eco-Nationalism: Anti-Nuclear Activism and National Identity in Russia, Lithuania, and Ukraine*. Durham: Duke University Press, 1996.

DeBardeleben, Joan. *The Environment and Marxism-Leninism*. Boulder: Westview Press, 1985.

DeBardeleben, Joan, and John Hannigan, eds. *Environmental Security and Quality after Communism: Eastern Europe and the Soviet Successor States*. Boulder: Westview Press, 1995.

Dixon, Jim. "Implementation of Agri-Environment Regulation 2078." In McCracken et al., *Farming on the Edge*.

Dominick, Raymond H., III. *The Environmental Movement in Germany: Prophets and Pioneers, 1871–1971.* Bloomington: Indiana University Press, 1992.

Doty, Roxanne Lynn. "Foreign Policy as Social Construction: A Post-Positivist Analysis of U.S. Counterinsurgency Policy in the Philippines." *International Studies Quarterly* 37 (1993): 297–320.

Dravnieks, A. *Latviešu literatūras vēsture.* Goppingen, Germany: Grāmatu draugs, 1946.

Dreifelds, Juris. "The Environmental Impact of Estonia, Latvia and Lithuania on the Baltic Sea Region." In DeBardeleben and Hannigan, *Environmental Security and Quality.*

———. "Two Latvian Dams: Two Confrontations." *Baltic Forum* 6, no. 1 (1989): 11–24.

Dundzila, Vilius Audrius. "Baltic Indigenous Religion." In *Encyclopedia of Religion and Nature,* edited by Bron Raymond Taylor. London: Continuum, 2005.

Dunsdorfs, Edgars. *Kārļa Ulmaņa dzīve: Ceļinieks, politiķis, diktators, moceklis.* Sweden: Daugava, 1978.

Eberhards, Guntis. "Daugavas programma." *Latvijas Kultūras fonda gaduraksti, 1987–1988.* Riga: Avots, 1991.

Ekmanis, Rolfs. *Latvian Literature under the Soviets, 1940–1975.* Belmont, MA: Nordland, 1978.

Eniņš, Guntis. *Koks–Dabas piemineklis.* Riga: Zinātne, 1982.

———. *Tepat Latvijā.* Riga: Liesma, 1984.

Escobar, Arturo. "Constructing Nature: Elements for a Poststructural Political Ecology." In Peet and Watts, *Liberation Ecologies.*

———. "Whose Knowledge, Whose Nature? Biodiversity, Conservation and the Political Ecology of Social Movements." *Journal of Political Ecology* 5 (1998): 53–82.

European Commission Directorate for Agriculture. *Summary Report: Agricultural Situation and Prospects in the Central and Eastern European Countries.* Luxembourg: Office for Official Publications of the European Communities, 1998.

European Commission Directorate General VI/A1. *Towards a Common Agricultural and Rural Policy for Europe.* Luxembourg: Office for Official Publications of the European Communities, 1997.

Eurostat. *Eurostat Yearbook 2000.* Luxembourg: Office for Official Publications of the European Communities, 2000.

Evans, Geoffrey, and Christine Lipsmeyer. "The Democratic Experience in Divided Societies: the Baltic States in Comparative Perspective." *Journal of Baltic Studies* 32, no. 4 (2001): 379–401.

Ezergailis, Andrievs. *The 1917 Revolution in Latvia.* Boulder: East European Monographs, 1974.

Feldman, Merje. "European Integration and the Discourse of National Identity in Estonia." *National Identities* 3, no. 1 (2001): 5–21.

———. "Justice in Space? The Restitution of Property Rights in Tallinn, Estonia." *Ecumene* 6, no. 2 (1999): 165–82.

Feshbach, Murray, and Alfred Friendly Jr. *Ecocide in the USSR: Health and Nature under Siege.* New York: Basic Books, 1992.

Fiedler, Peggy L., and Subodh K. Jain, eds. *Conservation Biology: The Theory and Practice of Nature Conservation, Preservation and Management.* New York: Chapman and Hall, 1992.

Fitzpatrick, Sheila. *Stalin's Peasants: Resistance and Survival in the Russian Village after Collectivization*. New York: Oxford University Press, 1994.

Forsyth, Tim. *Critical Political Ecology: The Politics of Environmental Science*. London: Routledge, 2003.

Franklin, Stuart. "Bialowieza Forest, Poland: Representation, Myth, and the Politics of Dispossession." *Environment and Planning A* 34 (2002): 1459–85.

Franzen, Jonathan. *The Corrections*. New York: Farrar, Straus and Giroux, 2000.

Frierson, Cathy A. *Peasant Icons: Representations of Rural People in Late Nineteenth Century Russia*. Oxford: Oxford University Press, 1993.

Gardner, Brian. *European Agriculture in the New Millenium*, vol. 2. London: Agra Europe, 1999.

Gellner, Ernest. "Nationalism in the Vacuum." In *Thinking Theoretically about Soviet Nationalities: History and Comparison in the Study of the USSR*, edited by Alexander J. Motyl. New York: Columbia University Press, 1992.

———. *Nations and Nationalism*. Ithaca: Cornell University Press, 1983.

Gillies, A. *Herder*. Oxford: Basil Blackwell, 1945.

Ginkel, John. "Identity Construction in Latvia's 'Singing Revolution': Why Inter-Ethnic Conflict Failed to Occur." *Nationalities Papers* 30, no. 3 (2002): 403–33.

Goldman, Marshall I. "Environmentalism and Nationalism: An Unlikely Twist in an Unlikely Direction." In Massey Stewart, *The Soviet Environment*.

Graham, Loren. *Science in Russia and the Soviet Union: A Short History*. Cambridge: Cambridge University Press, 1993.

Gray, Victor. "Identity and Democracy in the Baltics." *Democratization* 3, no. 2 (1996): 69–91.

Gruffud, Piers. "Remaking Wales: Nation-Building and the Geographical Imagination, 1925–50." *Political Geography* 14, no. 3 (1995): 219–39.

Gustafson, Thane. *Reform in Soviet Politics: Lessons of Recent Policies on Land and Water*. Cambridge: Cambridge University Press, 1981.

Haberer, Erich E. "Economic Modernization and Nationality in the Russian Baltic Provinces 1850–1900." *Canadian Review of Studies in Nationalism* 12, no. 1 (1985): 161–75.

Hamilton, F. E. Ian. "Transformation and Space in Central and Eastern Europe." *Geographical Journal* 165, no. 2 (1999): 135–44.

Hamm, Michael F., ed. *The City in Late Imperial Russia*. Bloomington: Indiana University Press, 1986.

Hasak-Lowy, Todd. "Thesis, Antithesis, Thesis: Nature in S. Yizhar's War of Independence Stories." In *Crisis and Memory: The Representation of Space in Modern Levantine Narrative*, edited by Ken Seigneurie. Wiesbaden: Reichert Verlag, 2003.

Hechter, Michael. *Internal Colonialism: the Celtic Fringe in British National Development, 1536–1966*. Berkeley: University of California Press, 1975.

Henriksson, Anders. "Riga: Growth, Conflict, and the Limitations of Good Government, 1850–1914." In Hamm, *The City in Late Imperial Russia*.

Hobsbawm, Eric J. *Nations and Nationalism since 1780*, 2nd ed. Cambridge: Cambridge University Press, 1990.

Hobsbawm, Eric, and Terence Ranger, eds. *The Invention of Tradition.* Cambridge: Cambridge University Press, 1983.

Hofer, Tamas. "The Creation of Ethnic Symbols from the Elements of Peasant Culture." In *Ethnic Diversity and Conflict in Eastern Europe,* edited by Peter F. Sugar. Oxford and Santa Barbara: ABC-Clio, 1980.

Hooson, David, ed. *Geography and National Identity.* Oxford: Blackwell, 1994.

Horowitz, Donald. *Ethnic Groups in Conflict.* Berkeley and Los Angeles: University of California Press, 1985.

Hroch, Miroslav. *Social Preconditions of National Revival in Europe: A Comparative Analysis of the Social Composition of Patriotic Groups among the Smaller European Nations,* trans. Ben Fowkes. Cambridge: Cambridge University Press, 1985.

Huntington, Samuel P. *The Clash of Civilizations and the Remaking of World Order.* New York: Simon & Schuster, 1996.

Ilbery, Brian, and Ian Bowler. "From Agricultural Productivism to Post-Productivism." In *The Geography of Rural Change,* edited by Ilbery. Edinburgh Gate, UK: Longman, 1998.

Ionescu, Ghita. "Eastern Europe." In Ionescu and Gellner, *Populism.*

Ionescu, Ghita, and Ernest Gellner, eds. *Populism: Its Meaning and National Characteristics.* New York: Macmillan, 1969.

Ishiyama, John T., and Marijke Breuning. *Ethnopolitics in the New Europe.* Boulder: Lynne Rienner, 1998.

IUCN. East Europe Programme. *The Wetlands of Central and Eastern Europe.* Gland, Switzerland: IUCN, 1993.

―――. *Guidelines for Protected Area Management Categories.* Gland, Switzerland: IUCN, 1994.

―――. *Parks for Life: Report of the IVth World Congress on National Parks and Protected Areas.* Gland, Switzerland: IUCN, 1993.

Īvāns, Dainis, and Artūrs Snips. *Domu Daugava, Sirdsdaugava.* Riga: Avots, 1989.

Jackson, Laura L. "The Role of Ecological Restoration in Conservation Biology." In Fiedler and Jain, *Conservation Biology.*

Jancar, Barbara. "The Environmental Attractor in the Former USSR: Ecology and Regional Change." In *The State and Social Power in Global Environmental Politics,* edited by Ronnie D. Lipschutz and Ken Conca. New York: Columbia University Press, 1993.

Jansons, Edvarts. "Dabas pieminekļi Latvija." In *Latvijas zeme, daba un tauta,* vol. II, edited by N. Malta and P. Galenieks. Rīga: Valters un Rapa, 1936.

Jarman, R. A. "Ecological Restoration: The End of Status Quo-ism in the National Trust?" *Biological Journal of the Linnean Society* 56, supplement (1995): 213–15.

Järve, Priit. "Two Waves of Language Laws in the Baltic States: Changes of Rationale?" *Journal of Baltic Studies* 33, no. 1 (2002).

Jaunsudrabiņš, Jānis. *Baltā grāmata.* Riga: Liesma, 1971.

―――. *Jaunsaimnieks un velns.* Stockholm: Daugava, 1949.

Jedlicki, Jerzy. "Polish Concepts of Native Culture." In Banac and Verdery, *National Character.*

Joravsky, David. *The Lysenko Affair.* Cambridge: Harvard University Press, 1970.

Jordan, William R., III. "'Sunflower Forest': Ecological Restoration as the Basis for a New Environmental Paradigm." In *Beyond Preservation: Restoring and Inventing Land-*

scapes, edited by A. Dwight Baldwin Jr., Judith De Luce, and Carl Pletsch. Minneapolis: University of Minnesota Press, 1994.

Josephson, Paul. *Totalitarian Science and Technology*. Atlantic Highlands, NJ: Humanities Press, 1996.

Jubulis, Mark. *Nationalism and Democratic Transition: The Politics of Citizenship and Language in Post-Soviet Latvia*. Lanham, MD: University Press of America, 2001.

Juska, Arunas. "Ethno-Political Transformation in the States of the Former USSR." *Ethnic and Racial Studies* 22, no. 3 (1999): 524–53.

Kaiser, Robert J. "Ethnoterritorial Conflict in the Former Soviet Union." In *The Challenge of Ethnic Conflict to National and International Order in the 1990s: Geographic Perspectives*. Washington, DC: United States Central Intelligence Agency Conference Report, 1995.

———. *The Geography of Nationalism in Russia and the USSR*. Princeton: Princeton University Press, 1994.

Kaldor, Mary. "Cosmopolitanism Versus Nationalism: The New Divide?" In *Europe's New Nationalism: States and Minorities in Conflict*, edited by Richard Caplan and John Feffer. New York: Oxford University Press, 1996.

Kalnins, Bruno. "The Social Democratic Movement in Latvia." In *Revolution and Politics in Russia: Essays in Memory of B. I. Nicolaevsky*, edited by Alexander and Janet Rabinowitch with Ladis K. D. Kristof. Bloomington: Indiana University Press, 1972.

Keller, Robert H., and Michael E. Turek. *American Indians and National Parks*. Tucson: University of Arizona Press, 1998.

Keohane, Robert O., and Marc Levy, eds. *Institutions for Environmental Aid*. Cambridge: MIT Press, 1996.

Kirby, David. *The Baltic World 1772–1993: Europe's Northern Periphery in an Age of Change*. London and New York: Longman, 1995.

Kiršteins, K. *Rokas grāmata mežkopjiem*. Vol. 3, Mežzinība. Riga: Mežu departaments, 1926.

Komarov, Boris. *The Destruction of Nature in the Soviet Union*. White Plains, NY: M. E. Sharpe, 1980.

Kronvalds, Atis. *Tagadnei: izlase*. Riga: Liesma, 1987.

Kundziņš, Pauls. *Latvju sēta*. Sweden: Daugava, 1974.

Kursīte, Janīna. *Mītiskais folklorā, literatūrā, mākslā*. Riga: Zinātne, 1999.

Lācis, Vilis. *Uz jauno krastu*. Riga: Latvijas valsts izdevniecība, 1958.

Lagerspetz, Mikko. "The Cross of Virgin Mary's Land: A Study in the Construction of Estonia's 'Return to Europe.'" *Finnish Review of East European Studies* 3–4 (1999): 17–28.

Laitin, David D. *Identity in Formation: The Russian-Speaking Populations in the Near Abroad*. Ithaca: Cornell University Press, 1998.

———. "Three Models of Integration and the Estonian/Russian Reality." *Journal of Baltic Studies* 34, no. 2 (2003): 197–222.

Lampland, Martha. *The Object of Labor: Commodification in Socialist Hungary*. Chicago: University of Chicago Press, 1995.

Langston, Nancy. "Forest Dreams, Forest Nightmares: An Environmental History of a Forest Health Crisis." In *American Forests: Nature, Culture, and Politics*, edited by Char Miller. Lawrence: University Press of Kansas, 1997.

Latvian Academy of Sciences Institute of Literature, Folklore, and Art. *Latviešu literatūres vēsture. 2. sējums: 1918–1945*. Riga: Zvaigzne ABC, 1999.

Latvian State Institute of Agrarian Economics. *Latvijas lauksaimniecība 1998: politika un attīstība*. Riga: Latvian State Institute of Agrarian Economics, 1999.

Lauristin, Marju, and Peeter Vihalemm. *Return to the Western World: Cultural and Political Perspectives on the Estonian Post-Communist Transition*. Tartu: Tartu University Press, 1997.

Lauristin, Marju, Peeter Vihalemm, and Mati Heidmets, eds. *The Challenge of the Russian Minority: Emerging Multicultural Democracy in Estonia*. Tartu: Tartu University Press, 2002.

Lazda, Paulis. "The Phenomenon of Russophilism in the Development of Latvian Nationalism in the Nineteenth Century." In *National Movements in the Baltic Countries during the Nineteenth Century*, edited by Aleksander Loit. Stockholm: Seventh Conference on Baltic Studies in Scandinavia, 1983.

Leopold, Aldo. "Wilderness." In *The Great New Wilderness Debate*, edited by J. Baird Callicott and Michael P. Nelson. Athens: University of Georgia Press, 1998.

Lewin, Moshe. *Russian Peasants and Soviet Power: A Study of Collectivization*. Evanston, IL: Northwestern University Press, 1968.

Liepins, Valdis. "Baltic Attitudes to Economic Recovery: A Survey of Public Opinion in the Baltic Countries." *Journal of Baltic Studies* 24, no. 2 (1993): 189–200.

Lieven, Anatol. *The Baltic Revolution: Estonia, Latvia, Lithuania and the Path to Independence*. New Haven: Yale University Press, 1993.

Lipschutz, Ronnie D., and Ken Conca, eds. *The State and Social Power in Global Environmental Politics*. New York: Columbia University Press, 1993.

Lowenthal, David. "European and English Landscapes as National Symbols." In Hooson, *Geography and National Identity*.

Lūsis, Kārlis, and Gints Šimanis, eds. *Krišjānim Valdemāram 170. Apceres, raksti, vēstules*. Riga: Latvijas Jūrniecības fonds, 1995.

Macnaghten, Phil, and John Urry. *Contested Natures*. London: Sage, 1998.

Malta, N., and P. Galenieks, eds. *Latvijas zeme, daba un tauta*. Vol. 2. Riga: Valters un Rapa, 1936.

Martin, Terry. *The Affirmative Action Empire: Nations and Nationalism in the Soviet Union, 1923–1939*. Ithaca: Cornell University Press, 2001.

Massey Stewart, John, ed. *The Soviet Environment: Problems, Policies and Politics*. Cambridge: Cambridge University Press, 1992.

McAfee, Kathleen. "Selling Nature to Save It? Biodiversity and Green Developmentalism." *Environment and Planning D: Society and Space* 17 (1999): 133–54.

McCarthy, James. "First World Political Ecology: Lessons from the Wise Use Movement." *Environment and Planning A* 24 (2002): 1281–1302.

McCracken, David I., Eric M. Bignal, and Susan E. Wenlock, eds. *Farming on the Edge: The Nature of Traditional Farmland in Europe*. Peterborough, UK: Joint Nature Conservation Committee, 1995.

McManus, Phil. "Histories of Forestry: Ideas, Networks and Silences." *Environment and History* 5, no. 2 (1999): 185–208.

Meadows, D. H., J. Randers, and W. W. Behrens. *The Limits to Growth*. London: Earth Island, 1972.

Medvedev, Zhores. *Soviet Agriculture*. New York: Norton, 1987.

Melderis, Kriss. *Mācība par mežu*. Riga: Valters un Rapa, 1939.

Melluma, Aija. *Latvijas PSR aizsargājamās dabas teritorijas*. Riga: Zinātne, 1979.

———. "Metamorphoses of Latvian Landscapes during Fifty Years of Soviet Rule." *Geojournal* 33, no. 1 (1994): 55–62.

Merķelis, Garlībs. *Izlase*. Riga: Liesma, 1969.

Merritt, Martha. "A Geopolitics of Identity: Drawing the Line between Russia and Estonia." *Nationalities Papers* 28, no. 2 (2000): 243–62.

Meyers, William H., and Natalija Kazlauskiene. "Land Reform in Estonia, Latvia, and Lithuania: A Comparative Analysis." In *Land Reform in the Former Soviet Union and Eastern Europe*, edited by Stephen K. Wegren. London and New York: Routledge, 1998.

Mežsaimniecības rakstu krājums, I sējums. Riga: Latvijas mežkopju un meža darbinieku biedrība, 1923.

Mežsaimniecības rakstu krājums, XV sējums. Riga: Latvijas mežkopju un meža darbinieku biedrība, 1937.

Mežsaimniecības rakstu krājums XVII. Riga: Latvijas Mežkopju savienība, 1939.

Millard, Frances. "Nationalism in Poland." In *Contemporary Nationalism in East Europe*, edited by Paul Latawski. New York: St. Martin's Press, 1995.

Ministry of Environmental Protection and Regional Development. *National Environmental Policy Plan for Latvia*. Riga: MEPRD and the Ministry of Housing, Physical Planning, and Environment of the Netherlands, 1995.

———. *National Report on Biological Diversity: Latvia*. Riga: MEPRD and United Nations Development Programme, 1998.

Misiunas, Romuald J., and Rein Taagepera. *The Baltic States: Years of Dependence, 1940–1990*, 2nd ed. Berkeley and Los Angeles: University of California Press, 1993.

Morley, David, and Kevin Robins. "No Place Like Heimat: Images of Home(land) in European Culture." In *Space and Place: Theories of Identity and Location*, edited by Erica Carter, James Donald, and Judith Squires. London: Lawrence & Wishart, 1993.

Motyl, Alexander J., ed. *Revolutions, Nations, Empires: Conceptual Limits and Theoretical Possibilities*. New York: Columbia University Press, 1999.

———. *Thinking Theoretically about Soviet Nationalities: History and Comparison in the Study of the USSR*. New York: Columbia University Press, 1992.

Muiznieks, Nils R. "The Daugavpils Hydro Station and Glasnost in Latvia." *Journal of Baltic Studies* 18 (Spring 1987): 63–70.

Nash, Roderick. *Wilderness and the American Mind*, third edition. New Haven: Yale University Press, 1982.

Neumann, Iver. *Russia and the Idea of Europe*. London: Routledge, 1996.

Neumann, Roderick P. "Ways of Seeing Africa: Colonial Recasting of African Society and Landscape in Serengeti National Park." *Ecumene* 2, no. 2 (1995): 149–69.

Norgaard, Ole, Lars Johannsen, Mette Skak, and René Hauge Sorensen. *The Baltic States after Independence*, 2nd ed. Cheltenham, UK: Edward Elgar, 1999.

Nove, Alec. *Soviet Agriculture: The Brezhnev Legacy and Gorbachev's Cure*. Los Angeles: Rand/UCLA Center for the Study of Soviet International Behavior, 1988.

Oelschlaeger, Max. *The Idea of Wilderness: From Prehistory to the Age of Ecology*. New Haven: Yale University Press, 1991.

Olwig, Kenneth R. "Reinventing Common Nature: Yosemite and Mount Rushmore–A Meandering Tale of a Double Nature." In Cronon, *Uncommon Ground*.

Pavlinek, Petr, and John Pickles. *Environmental Transitions: Transformation and Ecological Defence in Central and Eastern Europe*. London: Routledge, 2000.

Peet, Richard, and Michael Watts, eds. *Liberation Ecologies: Environment, Development, Social Movements*. London and New York: Routledge, 1996.

Peluso, Nancy Lee. *Rich Forests, Poor People: Resource Control and Resistance in Java*. Berkeley and Los Angeles: University of California Press, 1992.

Penrose, Jan. "Nations, States and Homelands: Territory and Territoriality in Nationalist Thought." *Nations and Nationalism* 8, no. 3 (2002): 277–97.

Peterson, D. J. *Troubled Lands: The Legacy of Soviet Environmental Destruction*. Boulder: Westview Press, 1993.

Pettai, Vello. "Emerging Ethnic Democracy in Estonia and Latvia." Presented at the annual meeting of the Association for the Advancement of Baltic Studies, 1994.

———. "The Games of Ethnopolitics in Latvia." *Post-Soviet Affairs* 12, no. 1 (1996): 40–50.

Pickett, Steward T. A., V. Thomas Parker, and Peggy L. Fiedler. "The New Paradigm in Ecology: Implications for Conservation Biology above the Species Level." In Fiedler and Jain, *Conservation Biology*.

Pickles, John, and Adrian Smith, eds. *Theorising Transition: The Political Economy of Post-Communist Transitions*. London: Routledge, 1998.

Plakans, Andrejs. "The Latvian National Awakening: Modernization of an Intellectual Milieu." *Bulletin of Baltic Studies* 1, no. 6 (1971): 9–13.

———. "The Latvians." In *Russification in the Baltic Provinces and Finland, 1855–1914*, edited by Edward C. Thaden. Princeton: Princeton University Press, 1981.

———. *The Latvians: A Short History*. Stanford: Hoover Institution Press, 1995.

———. "Peasants, Intellectuals, and Nationalism in the Russian Baltic Provinces, 1820–90." *Journal of Modern History* 46 (1974): 445–751.

Plotkin, Sidney. *Keep Out: The Struggle for Land Use Control*. Berkeley and Los Angeles: University of California Press, 1987.

Potter, Clive, and Philip Goodwin. "Agricultural Liberalization in the European Union: An Analysis of the Implications for Nature Conservation." *Journal of Rural Studies* 14, no. 3 (1998): 287–98.

Prazauskas, Algimantas. "The Influence of Ethnicity on the Foreign Policies of the Western Littoral States." In *National Identity and Ethnicity in Russia and the New States of Eurasia*, edited by Roman Szporluk. Armonk, NY: M. E. Sharpe, 1994.

Priedite, Aija. "Establishment of a Discourse about National Identity in Latvia in the Late 19th Century and the Early 20th Century." *Philosophical Discourse in Latvia: Humanities and Social Sciences. Latvia* 4, no. 25 (1999): 4–17.

Prieditis, Normunds. *Latvian Forests: Nature and Diversity*. Riga: WWF, 1999.

Priednieks, Jānis, Māris Kreilis, and Ilona Lodziņa. *National Biodiversity Action Plan for Latvia*. Riga: MEPRD, 1995.

Pryde, Philip R. "The Environmental Basis for Ethnic Unrest in the Baltic Republics." In Massey Stewart, *The Soviet Environment*.

Purapuķe, Jānis. *Savs kaktiņš–savs stūrītis zemes*. Lübeck: J. Šins, 1948.

Purs, Aldis. "Latvians as an Imagined Community." Presented at the Annual Convention of the Association for the Study of Nationalities, Columbia University, 1998.

Radkau, J. "Wood and Forestry in German History: In Quest of an Environmental Approach." *Environment and History* 2 (1996): 63–76.

Rainey, Thomas B. "Siberian Writers and the Struggle to Save Lake Baikal." *Environmental History Review* 15, no. 1 (1991): 48–60.

Ranger, Terence. *Voices from the Hills: Nature, Culture and History in the Matopos Hills of Zimbabwe*. Bloomington: Indiana University Press, 1999.

———. "Whose Heritage? The Case of the Matobo National Park." *Journal of Southern African Studies* 15, no. 2 (1989): 217–49.

Rasputin, Valentin. *Siberia on Fire*. DeKalb: Northern Illinois University Press, 1989.

Rauch, Georg von. *The Baltic States: The Years of Independence. Estonia, Latvia, Lithuania 1917–1940*, 2nd ed. New York: St. Martin's Press, 1995.

Reaka-Kudla, Marjorie L., Don E. Wilson, and Edward O. Wilson, eds. *Biodiversity II: Understanding and Protection Our Biological Resources*. Washington, DC: Joseph Henry Press, 1997.

Redclift, Michael. *Sustainable Development: Exploring the Contradictions*. London and New York: Methuen, 1987.

Richardson, Dick. "The Politics of Sustainable Development." In Baker et al., *Politics of Sustainable Development*.

Rikoon, J. Sanford. "Wild Horses and the Political Ecology of Nature Restoration in the Missouri Ozarks." Presented at the conference "Political Ecology at Home." Rutgers University Department of Geography, March 2003.

Rucavas Dabas Fonds. *Ir tāda vieta pie Baltijas jūras: Atklāsme par Rucavas pagastu*. Riga: WWF-Latvia and Rucavas Dabas Fonds, n.d.

Runte, Alfred. *National Parks: The American Experience*. Lincoln: University of Nebraska Press, 1979.

Sachs, Wolfgang, ed. *Global Ecology: A New Arena of Political Conflict*. London: Zed Books, 1995.

Saliņš, Staņislavs. *Latvijas dižkoki un retie koki*. Riga: Zinātne, 1974.

Sampson, Steven. "The Social Life of Projects: Importing Civil Society to Albania." In *Civil Society: Challenging Western Models*, edited by Chris Hann and Elizabeth Dunn. London: Routledge, 1996.

Sando, Paul. "Latvian Agriculture After Communism: Restructuring and Privatization." Ph.D. diss., Indiana State University, 1996.

Schama, Simon. *Landscape and Memory*. London: HarperCollins, 1995.

Schatz, Edward A. D. "Notes on the 'Dog That Didn't Bark': Eco-Internationalism in Late Soviet Kazakstan." *Ethnic and Racial Studies* 22, no. 1 (1999): 136–61.

Schwartz, Katrina. "'Masters in Our Native Place': The Politics of National Parks on the Road from Communism to 'Europe,'" *Political Geography* 25, no. 1 (January 2006): 42–71.

———. "Nature, Development, and National Identity: The Battle over Sustainable Forestry in Latvia," *Environmental Politics* 8, no. 3 (Autumn 1999): 99–118.

———. "'The Occupation of Beauty': Imagining Nature and Nation in Latvia," *East European Politics and Societies* (forthcoming).

———. *The Politics of Sustainable Forestry in Latvia: Property, Enterprise and the State in Transition from Communism.* Riga: WWF-International, 1996.

———. "Wild Horses in a 'European Wilderness': Imagining Sustainable Development in the Post-Soviet Countryside," *Cultural Geographies* 12, no. 3 (2005): 292–320.

Selwyn, Tom. "Landscapes of Liberation and Imprisonment: Towards an Anthropology of the Israeli Landscape." In *Anthropology of Landscape: Perspectives on Place and Space,* edited by Eric Hirsch and Michael O'Hanlon. Oxford: Clarendon Press, 1995.

Shafir, Gershon. "Relative Overdevelopment and Alternative Paths of Nationalism: A Comparative Study of Catalonia and the Baltic Republics." *Journal of Baltic Studies* 23, no. 2 (1992): 105–20.

Šilde, Ādolfs. *Latvijas vēsture 1914–1940.* Stockholm: Daugava, 1976.

Skujins, Zigmunds. "Imants Ziedonis Opens Clocks." Trans. Juris Silenieks. *World Literature Today* 72, no. 2 (1998): 297–300.

Slezkine, Yuri. "The USSR as a Communal Apartment, or How a Socialist State Promoted Ethnic Particularism." *Slavic Review* 53, no. 2 (1994): 414–52.

Slocock, Brian. "'Whatever Happened to the Environment?' Environmental Issues in the Eastern Enlargement of the European Union." In *Back to Europe: Central and Eastern Europe and the European Union,* edited by Karen Henderson. London: UCL Press, 1999.

Smith, Anthony D. *The Ethnic Origins of Nations.* Oxford and New York: Basil Blackwell, 1986.

———. *Myths and Memories of the Nation.* Oxford: Oxford University Press, 1999.

———. *The Nation in History: Historiographical Debates about Ethnicity and Nationalism.* Hanover, NH: University Press of New England, 2000.

———. *National Identity.* Reno: University of Nevada Press, 1991.

Smith, Graham. "The Ethnic Democracy Thesis and the Citizenship Question in Estonia and Latvia." *Nationalities Papers* 24, no. 2 (1996): 199–216.

———. "When Nations Challenge and Nations Rule: Estonia and Latvia as Ethnic Democracies." *International Politics* 33 (1996): 27–43.

Smith, Graham, Vivien Law, Andrew Wilson, Annette Bohr, and Edward Allworth. *Nation-Building in the Post-Soviet Borderlands: The Politics of National Identities.* Cambridge: Cambridge University Press, 1998.

Smurr, Rob. "Nationalizing Nature: The History, Preservation and Meaning of Glacial Erratic Boulders in Estonia." Presented at the annual meeting of the American Association for the Advancement of Slavic Studies, Pittsburgh, November 2002.

Spence, Mark David. *Dispossessing the Wilderness: Indian Removal and the Making of National Parks.* New York: Oxford University Press, 1999.

Staddon, Caedmon. "Localities, Natural Resources and Transition in Eastern Europe." *Geographical Journal* 165, no. 2 (1999): 200–08.

Stauter-Halsted, Keely. *The Nation in the Village: The Genesis of Peasant National Identity in Austrian Poland, 1848–1914.* Ithaca: Cornell University Press, 2001.

Stott, Philip, and Sian Sullivan. *Political Ecology: Science, Myth and Power.* London: Arnold, 2000.

Strijker, D., F. J. Sijtsma, and D. Wiersma. "Evaluation of Nature Conservation: An Application to the Dutch Ecological Network." *Environmental and Resource Economics* 16, no. 4 (2000): 363–78.

Strods, Heinrihs. *Latvijas Lauksaimniecības vēsture: no vissenākajiem laikiem lidz XX gs. 90. gadiem.* Riga: Zvaigzne, 1992.

Strods, Heinrihs, ed. *Latvijas mežu vēsture līdz 1940. gadam.* Riga: WWF, 1999.

Stucki, Erwin. "Balanced Development of the Countryside in Western Europe." *Nature and Environment,* no. 58. Strasbourg: Council of Europe Press, 1992.

Stukuls, Daina. "Imagining the Nation: Campaign Posters of the First Postcommunist Elections in Latvia." *East European Politics and Societies* 11, no. 1 (1997): 131–54.

Stukuls Eglitis, Daina. *Imagining the Nation: History, Modernity, and Revolution in Latvia.* University Park: Pennsylvania State University Press, 2002.

Suny, Ronald Grigor. *The Revenge of the Past: Nationalism, Revolution, and the Collapse of the Soviet Union.* Stanford: Stanford University Press, 1993.

Suško, Uvis, et al. *Natural Forests of Latvia: A Study on Biodiversity Structures, Dependent Species and Forest History.* Riga: WWF, 1997.

Švabe, Arveds. *Latvijas vēsture 1800–1914.* Uppsala: Daugava, 1962.

Swain, Nigel. "Getting Land in Central Europe." In *After Socialism: Land Reform and Social Change in Eastern Europe,* edited by Ray Abrahams. Providence, RI: Bergahn Books, 1996.

Szporluk, Roman. *Communism and Nationalism: Karl Marx versus Friedrich List.* Oxford: Oxford University Press, 1988.

Takacs, David. *The Idea of Biodiversity: Philosophies of Paradise.* Baltimore: Johns Hopkins University Press, 1996.

Tickle, Andrew, and Ian Welsh. "The 1989 Revolutions and Environmental Politics in Central and Eastern Europe." In Tickle and Welsh, *Environment and Society in Eastern Europe.*

Tickle, Andrew, and Ian Welsh, eds. *Environment and Society in Eastern Europe.* Harlow, UK: Longman, 1998.

Tisenkopfs, Talis. "Rurality as a Created Field: Towards an Integrated Rural Development in Latvia?" *Sociologia Ruralis* 39, no. 3 (1999): 411–30.

Tuan, Yi-Fu. "Perceptual and Cultural Geography: a Commentary." *Annals of the American Association of Geographers* 93, no. 4 (2003): 878–81.

Turner, Frederick. "Cultivating the American Garden: Toward a Secular View of Nature." *Harper's* 271, no. 8 (1985): 45–52.

———. "A Field Guide to the Synthetic Landscape: Toward a New Environmental Ethic." *Harper's* 276, no. 4 (1988): 49–55.

Turnock, David. "Sustainable Rural Tourism in the Romanian Carpathians." *Geographical Journal* 165, no. 2 (1999): 192–99.

United Nations Conference on Environment and Development. *Agenda 21 and the UNCED Proceedings*. New York: Oceana, 1992.

Unwin, Tim. "Contested Reconstruction of National Identities in Eastern Europe: Landscape Implications." *Norsk geografisk Tidsskrift* 53 (1999): 113–20.

———. "Rurality and the Construction of Nation in Estonia." In *Theorising Transition: The Political Economy of Post-Communist Transitions*, edited by John Pickles and Adrian Smith. London: Routledge, 1998.

Unwin, Tim, and Virginia Hewitt. "Banknotes and National Identity in Central and Eastern Europe." *Political Geography* 20 (2001): 1005–28.

Unwin, Tim, Judith Pallot, and Stuart Johnson. "Rural Change and Agriculture." In *East Central Europe and the Former Soviet Union: the Post-Socialist States*, edited by John Pickles and Adrian Smith. Harlow, UK: Pearson Education, 2004.

Upītis, H. *Rokas grāmata mežkopjiem, IV: Mežkopība*. Riga: Mežu departaments, 1939.

Valdemārs, Krišjānis. *Tēvzemei*. Riga: Avots, 1991.

Van Arkadie, Brian, and Mats Karlsson. *Economic Survey of the Baltic States*. New York: New York University Press, 1992.

Van Atta, Don, ed. *The "Farmer Threat": The Political Economy of Agrarian Reform in Post-Soviet Russia*. Boulder: Westview Press, 1993.

Van Wieren, S. E. "The Potential Role of Large Herbivores in Nature Conservation and Extensive Land Use in Europe." *Biological Journal of the Linnaean Society* 56, supplement (1995): 11–23.

Vera, Frans W. M. "Metaphors for the Wilderness: Oak, Hazel, Cattle and Horse." Ph.D. diss., Agricultural University, Wageningen, The Netherlands, 1997.

Verdery, Katherine. "The Elasticity of Land: Problems of Property Restitution in Transylvania." *Slavic Review* 53, no. 4 (1994): 1071–1108.

———. *National Ideology under Socialism: Identity and Cultural Politics in Ceausescu's Romania*. Berkeley: University of California Press, 1991.

———. "Nationalism and National Sentiment in Post-Socialist Romania." *Slavic Review* 52, no. 2 (1993): 179–203.

———. "Nationalism, Postsocialism, and Space in Eastern Europe." *Social Research* 63, no. 1 (1996): 77–95.

Vihalemm, Peeter. "Changing National Spaces in the Baltic Area." In *Return to the Western World: Cultural and Political Perspectives on the Estonian Post-Communist Transition*, edited by Marju Lauristin and Peeter Vihalemm. Tartu: Tartu University Press, 1997.

Wallis De Vries, Michiel F. "Large Herbivores and the Design of Large-Scale Nature Reserves in Western Europe." *Conservation Biology* 9, no. 1 (1995): 25–33.

Wanner, Catherine. *Burden of Dreams: History and Identity in Post-Soviet Ukraine*. University Park: Pennsylvania State University Press, 1998.

Wegren, Stephen K., ed. *Land Reform in the Former Soviet Union and Eastern Europe*. London: Routledge, 1998.

Weiner, Douglas R. *A Little Corner of Freedom: Russian Nature Protection from Stalin to Gorbachev*. Berkeley and Los Angeles: University of California Press, 1999.

———. *Models of Nature: Ecology, Conservation and Cultural Revolution in Soviet Russia.* Bloomington: Indiana University Press, 1988.

Wendt, Alexander. "Anarchy Is What States Make of It: The Social Construction of Power Politics." *International Organization* 46, no. 2 (1992): 391–441.

Whitby, Martin, ed. *The European Environment and CAP Reform: Policies and Prospects for Conservation.* Wallingford, UK: CAB International, 1996.

Williams, John Alexander. "'The Chords of the German Soul Are Tuned to Nature': The Movement to Preserve the Natural Heimat from the Kaiserreich to the Third Reich." *Central European History* 29, no. 3 (1996): 339–84.

Wilshusen, Peter R., Steven R. Brechin, Crystal L. Fortwangler, and Patrick C. West. "Reinventing a Square Wheel: Critique of a Resurgent 'Protection Paradigm' in International Biodiversity Conservation." *Society and Natural Resources* 15, no. 1 (2002): 17–40.

Wilson, Matthew A. "The Wolf in Yellowstone: Science, Symbol or Politics? Deconstructing the Conflict between Environmentalism and Wise Use." *Society & Natural Resources* 10, no. 5 (1997): 453–68.

Wolff, Larry. *Inventing Eastern Europe: The Map of Civilization on the Mind of the Enlightenment.* Stanford: Stanford University Press, 1994.

World Commission on Environment and Development. *Our Common Future: Report of the World Commission on Environment and Development.* Oxford: Oxford University Press, 1987.

World Resources Institute, World Conservation Union, and United Nations Environment Programme. *Global Biodiversity Strategy: Guidelines for Action to Save, Study, and Use Earth's Biotic Wealth Sustainably and Equitably.* Washington, DC: WRI, 1992.

World Wide Fund for Nature. *A History of WWF.* Gland, Switzerland: WWF, 1998.

———. *Latvia's Natural Heritage at the Crossroads.* Riga: WWF-Latvia, n.d.

———. *A New European Community Policy–Sustainable Regional Development.* Germany: WWF-Deutschland, 1997.

———. *World Conservation Strategy: Living Resource Conservation for Sustainable Development.* Gland, Switzerland: IUCN, UNEP, WWF, 1980.

———. *WWF Project 4568: Conservation Plan for Latvia, Final Report.* Riga: LU Ekoloģiskais centrs, 1992.

Wunderlich, Gene, ed. *Agricultural Landownership in Transitional Economies.* Lanham, MD: University Press of America, 1995.

Young, Craig, and Duncan Light. "Place, National Identity and Post-Socialist Transformations: An Introduction." *Political Geography* 20 (2001): 941–55.

Young, Crawford, ed. "The Dialectics of Cultural Pluralism: Concept and Reality." In Young, *The Rising Tide of Cultural Pluralism.*

———. *The Rising Tide of Cultural Pluralism: The Nation-State at Bay?* Madison: University of Wisconsin Press, 1993.

Zaslavsky, Victor. "Nationalism and Democratic Transition in Post-Communist Societies." *Daedalus* 121, no. 2 (1992): 97–121.

Ziedonis, Imants. *Kopotie rakst.* Vol. 9, *Tutepatās.* Riga: Nordik, 1998.

Ziegler, Charles E. *Environmental Policy in the USSR*. Amherst: University of Massachusetts Press, 1987.

———. "Political Participation, Nationalism and Environmental Politics in the USSR." In Massey Stewart, *The Soviet Environment*.